Australian **Wildflowers**

KEY GUIDE

Australian
Wildflowers

Envirobook

Leonard Cronin

Illustrated by
**Ruth Berry, Roslyn Devaux and
Marion Westmacott**

Leonard Cronin, is one of Australia's foremost natural history authors. Trained s a biologist, he is a prolific writer of books and articles on the Australian flora, fauna and the environment, bringing his own fascination with the natural world to the general reader. Among his other works are *The Australian Flora, The Australian Animal Atlas, Ancient Kingdoms and Natural Wonders, Koala, Presenting Australia's National Parks* and *The Illustrated Encyclopaedia of the Human Body.*

The illustrators of this volume, Marion Westmacott, Ruth Berry and Roslyn Devaux are leading Australian botanical artists.

First Published in Australia in 1987 by Reed Books
This edition first published in 2000 by Envirobook

Planned and produced by
Leonard Cronin Productions

Published in Australia by
Envirobook
38 Rose St, Annandale, NSW 2038

National Library of Australia Cataloguing-in-Publication Data:

Cronin, Leonard.
Key Guide to Australian Wildflowers

Bibliography
Includes index.

ISBN 0 85881 170 7

1. Wildflowers — Identification
1. Title (Series: Key Guides to Australian Flora)

582.13'0994

Publisher: Leonard Cronin
Design: Robert Taylor
Additional research: Gertrud Latour
Printer: Kyodo Printing Co, Singapore

Contents

Preface
to the second edition

Introduction

The first edition of *Key Guide to Australian Wildflowers* proved to be one of Australia's most popular field guides, going into numerous reprints. Since its first publication in 1987 taxonomists have been hard at work renaming and reclassifying species. This new, updated edition incorporates changes to the scientific names, new information about distribution, and new data where applicable.

Leonard Cronin, 2000.

Australians, and indeed, botanists throughout the world, have long been fascinated with the delicate beauty and adaptive ingenuity of our native wildflowers, often found flourishing in the most inhospitable of habitats, so far removed from the meadows and woodlands of the European pioneers.

Understandably so, for the isolation of this great island continent has allowed the evolutionary processes to develop a truly unique and diverse plant life.

Yet despite this interest the average nature lover has been sadly neglected, for among the whole array of books, journals and papers written about our native flora, few cater for the amateur observer who wants to identify and learn more, but finds the task of ploughing through complex botanical works or A to Zs of flowering plants too daunting. What so many people have been crying out for is an easy-to-use guide book that will enable the layperson to identify an unknown wildflower simply, and without any previous botanical knowledge.

Botanists use key systems to accurately identify unknown specimens, but these are traditionally difficult to use and require an extensive knowledge of botanical terms. To overcome this problem I have devised a very simple key specifically for use with this book, relying on easily identifiable visual characteristics of the plants described.

This has meant transcending the boundaries of traditional botanical groupings where plants are listed together in their families. Instead, I have grouped the plants according to their flower shapes. Thus you will find all the bell-shaped flowers gathered together, all the pea-shaped flowers together, all the daisy-like flowers together, and so on. The easy-access visual key at the beginning of the book directs you to the appropriate page, where it is a simple matter to identify the flower accurately and read a concise description of the species.

Essential to the book are the magnificent colour illustrations by Ruth Berry, Roslyn Devaux and Marion Westmacott. Their style and accuracy emphasise the characteristics of the plants, and add a visual appeal to the book that photographs could never hope to match.

600 wildflowers are illustrated and described. They have been carefully selected to represent the most commonly encountered wildflowers in Australia. They grow in areas which are easily accessible to the average nature lover, including picnic areas, National Parks and along the roadsides.

The botanical and common names of each plant are given, together with other essential details, such as colour variations, size, leaf shape, flowering season, fruit and habitat. The descriptions are as non-technical as possible, and the few botanical terms used are carefully explained in the comprehensive glossary at the end of the book.

This volume is the culmination of more than two years of concentrated research, writing and revising; and reading the plant descriptions you will notice the wide variety of sizes, shapes and flowering times given for those species found in a wide range of habitats. This in itself caused many problems for early botanists who often thought they were dealing with quite different species, hence the variety of botanical names given to some plants by different botanists. This is indicated by the use of the word synonym, abbreviated to syn. in the text. The names in parentheses are alternatives still in use, although strictly speaking the first Latin name given is the correct one. But the naming of Australian plants is an ongoing process, and botanists are still describing new species and renaming others.

I think it is appropriate that a plea be made here for greater conservation of our wilderness areas. We have, in Australia, one of the greatest storehouses of biological diversity left on the planet. It has taken many millions of years to create such genetic variation, and by the destruction of "unprofitable" bush land and rainforests we are removing the habitats of species that have never even been described, let alone examined for their place in the web of life. Nature is an intertwined and interdependent system, and biologists are more and more coming to realise the danger of destroying the habitats of thousands of species for the cultivation of the few species considered to be of economic importance.

Using this book as a guide you will be surprised how quickly you will become familiar with many new plants and confident to approach other botanical books for further information. The benefits of identifying the plants around us are enormous, the eye is easily trained to pick out known characteristics, and the bush becomes at once familiar, and a new, fascinating and infinitely varied world opens up before us.

How to use this guide

This book is designed to enable you to identify the common Australian wildflowers quickly and easily. To make this as simple as possible flowers with similar shapes have been grouped together regardless of their family or genus relationships. The simple visual key refers you to individual pages in the book, with six species illustrated on the pages opposite.

Many wildflowers belonging to the same family do, of course, have similarly-shaped flowers, and these appear on the same plate. Others in the same family may look completely different, and these I have grouped with similarly-shaped flowers of different families.

To use this book you do not need any knowledge of plant classification. Simply look at the specimen you wish to identify and compare its shape with the generalised flowers shown in the left-hand columns of the **Key Guide**. Having found the nearest corresponding **General flower shape**, look at the right-hand column to find the nearest **Typical flower shape**. Now simply turn to the pages indicated and identify your specimen by looking at the illustrations and descriptions given.

Example

In open forest country along the coast of New South Wales you come across a stiff shrub with large, yellow, open flowers and 5-lobed, broad leaves.

1. Using the **Key Guide** you find, by looking down the left-hand columns, that the **Open** flower shape resembles the generalised shape of your specimen.

2. In the right-hand column you will see that your specimen corresponds to the **Typical flower shape** with 5 overlapping lobes, and that you are referred to pages 36, 40 and 164-176.

3. On page 169 you will find that the the illustration of the Yellow Hibiscus matches your specimen. Confirm your identification and discover more about the plant by reading the description of the Yellow Hibiscus (*Hibiscus diversifolius*) opposite.

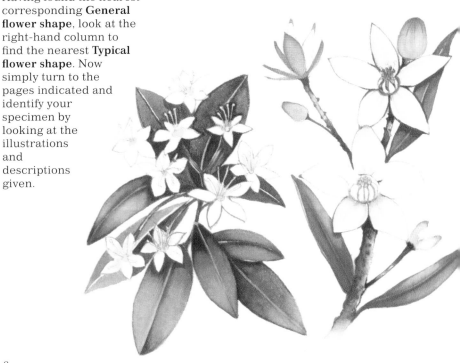

Key Guide

General flower shape	Typical flower shape	Page

	3-lobed	132, 134, 184
	4-lobed	136-144
Open	5 separate lobes	38, 44, 144-166
	5 overlapping lobes	36, 40, 164-176
	6 lobes	176-184
	8+ lobes	184-188
Composite	Buttons	80-84
	Daisy-like	84-92

Key Guide

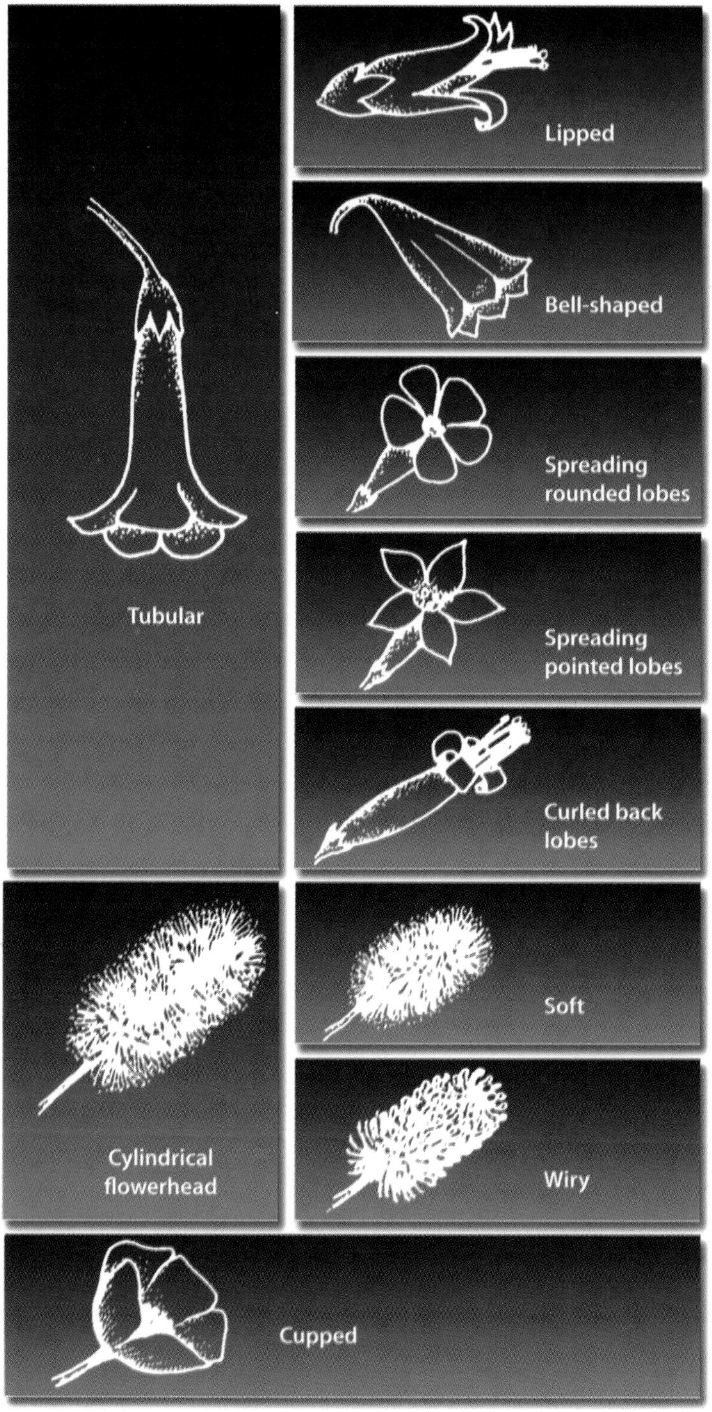

General flower shape	Typical flower shape	Page

Lipped — 12-24

Bell-shaped — 24-30

Spreading rounded lobes — 42-46

Spreading pointed lobes — 48-62

Curled back lobes — 62-66

Tubular

Soft — 94-102

Wiry — 102-106

Cylindrical flowerhead

Cupped — 30-40, 146, 148

Key Guide

General flower shape	Typical flower shape	Page
Orchid	Tailed	120, 122
	Hooded	122-126
	Regular spreading lobes	126-132
Pea	Purple	188-194
	Red	196, 198
	Yellow	200-210
Globular flowerhead		68-78, 162
Spidery flowerhead		108-118

Eremophila alternifolia
Magenta Emu Bush

Erect, slender shrub to 3 m high, with rough, slightly sticky branches. **Leaves** Alternate, crowded towards the ends of the branches, fleshy, narrow-linear, 10-55 mm long and 1-2 mm wide, usually with a curved, pointed tip. **Flowers** Rosy-red, rarely white or yellow, sometimes with red spots inside and out, tubular, 18-35 mm long with a single, broad, lower lip and a 4-lobed upper lip, with 4 long stamens. They are solitary on axillary stalks 1-3 cm long. **Fruits** Woody ovoid drupes 5-7 mm long. **Flowering** Spring and summer. **Habitat** Widespread on dry rocky hills and sandy plains in WA, NT, SA and NSW. **Family** Myoporaceae.

Eremophila calorhabdos
Red Rod. Spiked Eremophila

Slender erect shrub to 2 m high. **Leaves** Alternate, crowded, lanceolate to ovate-oblong with finely toothed margins, 15-25 mm long and 5-10 mm wide, slightly concave, initially covered in short hairs. **Flowers** Pink to purplish-red, tubular, 2-3 cm long including 4 long protruding stamens, with a single narrow lower lip and a 4-lobed upper lip. They are solitary in the leaf axils, often forming a leafy spike. **Fruits** Ovoid drupes about 7 mm long. **Flowering** Spring and summer. **Habitat** Open woodlands and semi-arid areas of inland southern WA. **Family** Myoporaceae.

Eremophila duttonii
Budda. Emu Bush. Harlequin Fuchsia Bush

Compact, rounded shrub to 4 m high. **Leaves** Alternate, crowded towards the ends of the branches, linear to narrow-lanceolate, 15-60 mm long and 3-5 mm wide. **Flowers** Red outside and yellow inside, tubular, 25-35 mm long with a single, narrow, lower lip and a 4-lobed upper lip, with 4 protruding stamens. They are solitary on long axillary stalks. **Fruits** Woody, broad, ovoid drupes to 11 mm long. **Flowering** Late winter and spring. **Habitat** Widespread on dry red-soils, sandplains and rocky hills, inland in all mainland states except Vic. **Family** Myoporaceae.

Eremophila glabra
Common Emu Bush. Tar Bush

Erect, prostrate or spreading shrub to 3 m high. Young growth is often sticky with white, felty hairs. **Leaves** Alternate, lanceolate to elliptic, sometimes with toothed margins, 1-6 cm long and 2-18 mm wide. **Flowers** Red, pink or yellow-green, tubular, 20-30 mm long, hairy inside and sometimes outside, with a curled, narrow, lower lip and a 4-lobed, curled, upper lip, with 4 protruding stamens. They are solitary on short axillary stalks. **Fruits** Dry or fleshy ovoid drupes 4-10 mm long. **Flowering** Mostly from winter to summer. **Habitat** Widespread, often with mallee and Bimble Box, inland in all mainland states. **Family** Myoporaceae.

Eremophila latrobei
Crimson Turkey Bush

Erect, spreading shrub to 3 or 4 m high, often covered in white, downy hairs. **Leaves** Alternate, linear to lanceolate or oblong, broader towards the tips, usually with curled-under margins, 1-9 cm long and 1-6 mm wide, often hairy. **Flowers** Red to purplish-pink, rarely white or yellow, tubular, 12-35 mm long with a broad lower lip and a 4-lobed upper lip, with 4 protruding stamens, solitary on axillary stalks about 1 cm long. **Fruits** Woody ovoid drupes, 5-10 mm long. **Flowering** Winter and spring. **Habitat** Semi-arid areas in Bimble Box and Mulga country, inland in all mainland states except Vic. **Family** Myoporaceae.

Lechenaultia hirsuta
Scarlet or Hairy Lechenaultia

Sprawling, perennial herb to 30 cm high with wiry-hairy stems. **Leaves** Narrow-linear, pointed, hairy, 1-3 cm long. **Flowers** Scarlet, tubular, 25-30 mm long, with 3 wedge-shaped lower lips and a 2-lobed, erect, upper lip, with 5 stamens. They are solitary on long axillary stalks. **Fruits** 4-5 valved, dry, woody pods. **Flowering** Spring and summer. **Habitat** Sand heaths along the central west coast of WA. **Family** Goodeniaceae.

Eremophila alternifolia

Eremophila latrobei

Eremophila calorhabdos

Eremophila duttonii

Eremophila glabra

Lechenaultia hirsuta

Prostanthera aspalathoides
Scarlet Mint Bush

Low, spreading, rigid and hairy aromatic shrub to 1 m high. **Leaves** Opposite, crowded, linear to needle-like, thick and channelled above, aromatic, 1-7 mm long and about 1 mm wide. **Flowers** Bright-red, to pink-red, orange or rarely yellow, downy, tubular, 1-2 cm long with a 3-lobed lower lip and erect, 2-lobed upper lip, with 4 stamens. They are solitary on short axillary stalks. **Fruits** Small nuts. **Flowering** Spring. **Habitat** Sandy soils in moderately dry areas among mallee and other shrubs in central NSW, Qld, southwestern Vic. and southern SA. **Family** Lamiaceae.

Chloanthes stoechadis

Erect or straggling woolly shrub to 1 m high. **Leaves** Opposite in whorls of 3, fleshy, covered with nodules, narrow-linear to lanceolate, 1-5 cm long and 1-5 mm wide with curled-under margins, rough above with white, woolly hairs below. **Flowers** Green-blue to green-yellow, tubular, 18-45 mm long with a 2-lobed upper lip and 3-lobed lower lip, hairy outside the base of the tube, with 4 protruding stamens. They are solitary in the leaf axils and almost stalkless. **Fruits** Elliptic, 2-valved capsules, 3-5 mm long. **Flowering** Most of the year. **Habitat** Widespread in sandy heaths and dry sclerophyll forests along the central and southern coast and tablelands of NSW. **Family** Verbenaceae.

Eremophila maculata
Spotted Emu Bush. Native Fuchsia

Spreading, bushy shrub to 3 m high, with downy branches. **Leaves** Alternate, elliptic to linear or lanceolate, pointed, 5-50 mm long and 2-14 mm wide. **Flowers** Pink to red, occasionally yellow, spotted inside, tubular, 20-35 mm long with a long, curled lower lip and 4-lobed upper lip, with 4 protruding stamens. They are solitary on curved axillary stalks 10-25 mm long. **Fruits** Green to purple ovoid fleshy drupes, 12-20 mm long. **Flowering** From autumn to summer. **Habitat** Widespread on clay soils, inland in all mainland states. **Family** Myoporaceae.

Philydrum lanuginosum
Frogsmouth. Woolly Waterlily

Erect, perennial, aquatic herb to 1 m high, covered with grey woolly hairs. **Leaves** Mostly crowded, sword-shaped, arising from and sheathing the base of the plant, 20-60 cm long, with smaller floral leaves along the stem. **Flowers** Yellow, tubular, about 12 mm long and 10 mm across with prominent upper and lower lips; hairy outside, with 2 central lobes sheathing a single stamen. They are stalkless in the axis of leafy bracts, forming a terminal spike. **Fruits** 3-valved oblong capsules about 12 mm long. **Flowering** Summer. **Habitat** Widespread in swampy sites of northern WA, NT, Qld, NSW and Vic. **Family** Philydraceae.

Conospermum mitchellii
Victorian Smoke Bush

Erect shrub with few slender stems to 3 m high. **Leaves** Crowded, alternate and erect, stiff, narrow-linear, 6-15 mm long and 1-4 mm wide. **Flowers** Creamy-white, sometimes blue-tipped, tubular, 5-8 mm long, hairy outside, with a 3-lobed lower lip and curled upper lip, with 4 stamens. They are stalkless and crowded in flat-topped terminal clusters 10-15 cm across. **Fruits** Dry, hairy, single-seeded, 2-3 mm long. **Flowering** Spring. **Habitat** Sandy soils in heaths and open forests in Vic. **Family** Proteaceae.

Euphrasia glacialis
Glacial Eye Bright

Stout, perennial herb to 12 cm high, parasitic on the roots of other plants, sometimes regarded as an alpine form of *E. collina*. **Leaves** Opposite, stalkless, elliptic to wedge-shaped, coarsely toothed, pale-green, hairy below, 5-9 mm long and 4-6 mm wide. **Flowers** White with a yellow patch on the lower lobe, tubular, 10-12 mm long, with a 2-lobed, curled upper lip and spreading, 3-lobed lower lip, with 4 stamens. They are stalkless in loose terminal clusters. **Fruits** Dry flattened capsules. **Flowering** Spring and summer. **Habitat** Open alpine moors in the southern tablelands of NSW. **Family** Scrophulariaceae.

Chloanthes stoechadis

*Eremophila
maculata*

*Euphrasia
glacialis*

*Conospermum
mitchellii*

*Prostanthera
aspalathoides*

Philydrum lanuginosum

Prostanthera cuneata
Alpine Mint Bush

Compact, dense, aromatic shrub to 1 m high, with hairy branches. **Leaves** Opposite, crowded, obovate to orbicular, almost stalkless, strongly aromatic, 4-7 mm long and 3-5 mm wide, glossy dark-green above and paler below. **Flowers** White to pale-mauve with purple and yellow spots inside, tubular, 10-15 mm long with a 3-lobed, spreading lower lip and erect, 2-lobed upper lip, with 4 stamens. They are solitary in the upper leaf axils, sometimes forming leafy racemes. **Flowering** From spring to autumn. **Habitat** Exposed rocky sites in heaths and shrublands in alpine and sub-alpine areas, often under Snowgums in NSW, Vic. and Tas. **Family** Lamiaceae.

Prostanthera lasianthos
Christmas Mint Bush

Upright aromatic shrub, occasionally a small tree, 1-8 m high. **Leaves** Opposite, lanceolate to oblong, often with toothed margins, 4-12 cm long and 10-35 mm wide, paler-green below. **Flowers** White, occasionally mauve, with orange and purple spots inside, tubular, downy inside and out, 10-15 mm long with a 3-lobed, spreading, lower lip and erect, 2-lobed upper lip. They are solitary or in short leafless racemes. **Flowering** From spring to autumn. **Habitat** Lowland rainforests, sclerophyll forests and sub-alpine woodlands in Qld, NSW, Vic. and Tas. **Family** Lamiaceae.

Prostanthera nivea
Snowy Mint Bush

Upright shrub to 4 m high. **Leaves** Opposite, stalkless, narrow-linear to narrow-lanceolate, sometimes concave, 1-5 cm long and 1-3 mm wide, smooth or with dense white hairs. **Flowers** White, sometimes mauve-tinged with orange/yellow spots inside, hairy, tubular, 10-18 mm long with a 3-lobed spreading lower lip and erect 2-lobed upper lip. They are solitary in the upper leaf axils forming interrupted leafy racemes. **Flowering** Spring. **Habitat** Widespread in heaths, forests and woodlands on the coast and inland in Qld, NSW and Vic. **Family** Lamiaceae.

Prostanthera striatiflora
Striped Mint Bush

Upright aromatic shrub to 2 m high. **Leaves** Opposite, concave, oblong to lanceolate or narrow-ovate, 8-30 mm long and 2-10 mm wide. **Flowers** White with purple streaks and yellow spots, tubular, hairless or slightly downy outside, 10-25 mm long with a 3-lobed, spreading lower lip and erect, 2-lobed upper lip. They are solitary in the upper leaf axils forming crowded leafy racemes. **Flowering** Winter and spring. **Habitat** Rocky sites in open woodlands inland in NSW, WA, SA and NT. **Family** Lamiaceae.

Westringia fruticosa
Coast Rosemary

Compact aromatic shrub to 2 m high. **Leaves** Leathery, in whorls of 3-5, crowded, linear to narrow-lanceolate with curved-back margins, hairy below, 10-30 mm long and 2-15 mm wide. **Flowers** White or pale-mauve with orange or purplish spots inside, tubular, 10-15 mm long with a 3-lobed, spreading lower lip and erect upper lip, solitary in the upper leaf axils. **Flowering** Most of the year. **Habitat** Heaths and windswept headlands of central and southern coastal NSW. **Family** Lamiaceae.

Hemiandra pungens
Snake Bush

Spreading or prostrate aromatic shrub to 1 m high. **Leaves** Stalkless, fleshy, stiff, pointed, narrow-linear, flat or concave, about 12 mm long. **Flowers** White to mauve-pink with red spots inside, tubular, variable but often about 2 cm across with a 3-lobed, spreading lower lip and erect upper lip, arranged in small axillary racemes. **Flowering** Spring and summer. **Habitat** Widespread in sandy heaths along the coast of southwestern WA. **Family** Lamiaceae.

*Prostanthera
lasianthos*

*Hemiandra
pungens*

*Prostanthera
striatiflora*

*Westringia
fruticosa*

*Prostanthera
cuneata*

*Prostanthera
nivea*

Prostanthera denticulata
Rough Mint Bush

Straggling, aromatic, slender shrub to 1 m high and 2 m wide with hairy branches. **Leaves** Opposite, broad-lanceolate to narrow-ovate with curled-under margins, almost stalkless, 5-12 mm long and about 1 mm wide with rough upper surfaces. **Flowers** Purple, tubular, 7-12 mm long with a 3-lobed lower lip and erect, 2-lobed upper lip, arranged in pairs in slender leafy racemes. **Flowering** Spring and early summer. **Habitat** Scattered in damp sites in sandy and gravelly soils in sclerophyll forests and woodlands from southeastern Qld to NSW and Vic. **Family** Lamiaceae.

Prostanthera melissifolia
Balm Mint Bush

Slender, aromatic, upright shrub to 3 m high with hairy branches. **Leaves** Opposite, ovate, toothed, thin, dull-green above and paler below, 15-50 mm long and 1-2 cm wide. **Flowers** Pink to violet, tubular, 8-15 mm long with a 3-lobed, spreading, lower lip and erect, 2-lobed upper lip. They are arranged in loose, leafless racemes. **Flowering** Spring. **Habitat** Damp forested slopes and stream banks in restricted areas of southeastern NSW and southern Vic. **Family** Lamiaceae.

Prostanthera ovalifolia

Erect, bushy shrub to 4 m high. **Leaves** Opposite, thick and flat, ovate to oblong or lanceolate, soft, dark-green above and paler below, 1-4 cm long and 3-20 mm wide. **Flowers** Mauve to deep purplish-blue, rarely white, tubular, 6-10 mm long with a 3-lobed, spreading lower lip and erect, 2-lobed upper lip, slightly-hairy outside. They are solitary or in short terminal racemes. **Flowering** Late winter and spring. **Habitat** Widespread in rocky sites in sclerophyll forests along the coast and inland in Qld and NSW. **Family** Lamiaceae.

Prostanthera rotundifolia
Round-leaved Mint Bush

Upright, aromatic, hairy, bushy shrub to 3 m high. **Leaves** Opposite on long stalks, spathulate to broadly-ovate or orbicular, entire or with rounded teeth, thick, dark-green above and paler below, 4-15 mm long and 4-14 mm wide. **Flowers** Purple to blue, sometimes pinkish, tubular, 10-15 mm long with a 3-lobed, spreading, lower lip and erect, 2-lobed upper lip. They are solitary in short, leafy, terminal racemes. **Flowering** Spring. **Habitat** Widespread in sheltered rocky sites in sclerophyll forests and woodlands of the coast and tablelands of NSW, Vic. and Tas. **Family** Lamiaceae.

Prostanthera scutellarioides

Erect or spreading, slightly aromatic shrub, 0.3-2.5 m high. **Leaves** Opposite, linear to lanceolate with curved-back margins, 6-25 mm long and 1-2 mm wide. **Flowers** Purple, tubular, 7-10 mm long with a 3-lobed, spreading, lower lip and erect, 2-lobed upper lip. They are solitary or in axillary racemes on short stalks. **Flowering** Spring and early summer. **Habitat** Widespread in dry sclerophyll forests and woodlands along the coast and table-lands of Qld and NSW. **Family** Lamiaceae.

Pratia purpurascens
White Root

Prostrate or creeping perennial herb. **Leaves** Alternate, ovate to lanceolate, irregularly-toothed, 10-25 mm long and 4-10 mm wide, purplish below. **Flowers** White, pale-pink to lilac, tubular, 8-10 mm long, with a 3-lobed, spreading lower lip and erect, 2-lobed upper lip. They are solitary on slender axillary stalks. **Fruits** 2-valved capsules, 3-10 mm long. **Flowering** From spring to autumn. **Habitat** Widespread in damp shady sites and wet sclerophyll forests of the coast and tablelands in Qld, NSW, southeastern Vic. and southeastern SA. **Family** Campanulaceae.

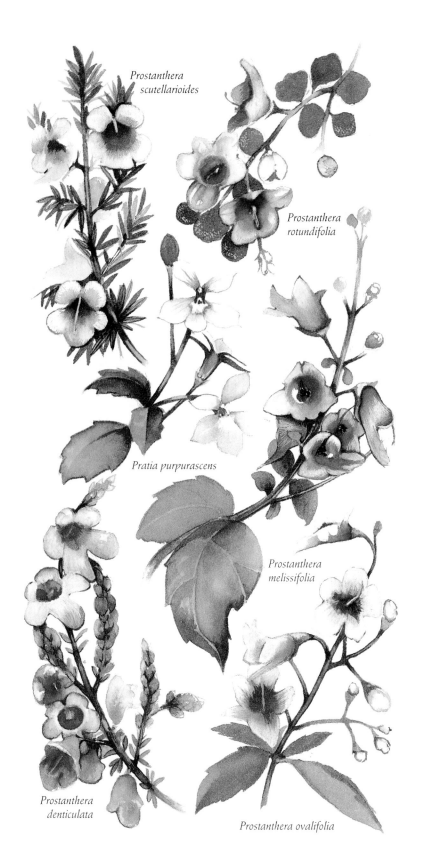

*Prostanthera
scutellarioides*

*Prostanthera
rotundifolia*

Pratia purpurascens

*Prostanthera
melissifolia*

*Prostanthera
denticulata*

Prostanthera ovalifolia

Hemigenia purpurea

Erect, slender shrub to 2 m high, with hairy branches. **Leaves** Linear to needle-like, thick and fleshy, channelled above, in whorls of 3-4, 8-16 mm long and 1-2 mm wide. **Flowers** Pale-purple to blue, hairy outside, tubular, 6-10 mm long with a long, 3-lobed, spreading lower lip and erect upper lip. They are solitary in the upper leaf axils. **Flowering** Most of the year. **Habitat** Sandy heaths of the coast and tablelands in central eastern and southeastern NSW. **Family** Lamiaceae.

Westringia eremicola   Slender Westringia

Upright shrub to 2 m high. **Leaves** Linear to narrow-elliptic, often with curled-under margins, hairy, in whorls of 3-4, 7-30 mm long and 1-2 mm wide. **Flowers** Pale-violet to purple or white, with brownish spots inside, hairy, tubular, 5-9 mm long with a long, 3-lobed, spreading lower lip and erect, 2-lobed upper lip. They are solitary in the leaf axils on very short stalks. **Flowering** Year round. **Habitat** Widespread in rocky or sandy sites of the coast and tablelands of Qld, NSW, Vic. and SA. **Family** Lamiaceae.

Westringia glabra   Violet Westringia

Upright, variable shrub, 0.5-5 m high. **Leaves** Elliptic to lanceolate, in whorls of 3-4, thin, paler below, flat or with slightly curved-back margins, 1-5 cm long and 2-15 mm wide. **Flowers** Mauve with orange spots inside, tubular, 6-10 mm long, with a long, 3-lobed lower lip and erect, 2-lobed upper lip. They are solitary on very short stalks or in small clusters in the upper leaf axils. **Flowering** Year round. **Habitat** Low, open woodlands and shrublands in shallow soils and rocky gorges near streams in southern coastal Qld, the northern tablelands of NSW and Vic. **Family** Lamiaceae.

Ajuga australis   Australian Bugle

Erect or spreading, downy, perennial herb to 60 cm high. **Leaves** Opposite in a rosette, ovate to oblong, entire or coarsely-toothed, hairy, soft, 4-12 cm long. **Flowers** Blue or purple, rarely pinkish, tubular, 7-20 mm long with a long, 3-lobed lower lip, and inconspicuous, 2-lobed upper lip. They are stalkless in whorls of 6-20 in the leaf axils, often forming a leafy spike. **Flowering** Spring and summer. **Habitat** Sandy soils and rocky outcrops in open forests of the coast and inland in all states except WA and NT. **Family** Lamiaceae.

Chloanthes parviflora

Upright or straggling, woolly shrub, to 1 m high. **Leaves** Pale-green, opposite in whorls of 3, fleshy, stalkless, covered with nodules, narrow-linear to lanceolate, 1-4 cm long and 2-5 mm wide with curled-under margins, rough above with white, woolly hairs below. **Flowers** Pale-blue or lavender, spotted orange-brown inside, tubular, 15-32 mm long with a large, hairy, lower lip and 4-lobed upper lip, hairy inside the tube with 4 protruding stamens. They are solitary in the leaf axils and almost stalkless. **Flowering** Most of the year. **Habitat** Widespread in sandy heaths and dry sclerophyll forests in Qld and NSW. **Family** Verbenaceae.

Euphrasia collina   Eye Bright

Upright perennial herb to 80 cm high, parasitic on the roots of other plants. **Leaves** Opposite, stalkless, sometimes with 1-6 pairs of coarse teeth, oblong to elliptical, 5-19 mm long and 2-9 mm wide. **Flowers** White, mauve or violet, sometimes with white extremities, tubular, 12-20 mm long with a curled, 2-lobed upper lip and spreading, 3-lobed lower lip, with 4 stamens. They are stalkless in terminal spikes of 20-50 flowers. **Fruits** Oblong, compressed capsules. **Flowering** Spring and summer. **Habitat** Widespread in moist sites of the coast and tablelands in southern Qld, NSW, Vic., SA, southwestern WA and Tas. **Family** Scrophulariaceae.

*Chloanthes
parviflora*

Hemigenia purpurea

*Euphrasia
collina*

Westringia eremicola

*Westringia
glabra*

Ajuga australis

Utricularia dichotoma
Fairy Aprons. Purple Bladderwort

Erect, slender, carnivorous herb, to 30 cm high. **Leaves** Mainly submerged or subterranean, finely-divided and highly modified to trap minute aquatic animals in small, hollow bladders. Emerged leaves arise from the roots, linear to ovate, 2-40 mm long and usually absent at flowering time. **Flowers** Purple to lilac or white with a yellowish throat, tubular, 12-20 mm long, spurred at the base, with a large lower lip. They are terminal, solitary or in small clusters on long stems. **Flowering** From spring to autumn. **Habitat** Peaty margins of heathland swamps and alpine bogs in all states except NT. **Family** Lentibulariaceae.

Hybanthus monopetalus
Slender Violet-bush

Upright perennial herb to 60 cm high with slender, wiry stems. **Leaves** Linear to oblong or narrow-elliptic with curled margins, alternate on the lower stem and opposite higher up, 1-9 cm long. **Flowers** Blue, shortly tubular, 13-20 mm long with a spade-shaped, large lower lip and 4 small upper petals, with 5 stamens. Arranged in small leafless racemes on long axillary stalks. **Fruits** 3-valved globular capsules, 3-6 mm long. **Flowering** Spring and summer. **Habitat** Widespread on sandy soils or rocky outcrops in dry forests and woodlands in Qld, NSW, SA and Vic. **Family** Violaceae.

Hybanthus vernonii
Erect Violet

Upright perennial herb to 1 m high with slender wiry stems. **Leaves** Alternate, linear to narrow-ovate or lanceolate, 5-45 mm long. **Flowers** Mauve, shortly tubular, 6-13 mm long, with a spade-shaped, large lower lip and 4 small, curled upper petals, with 5 stamens, arranged in small clusters in the upper leaf axils. **Fruits** 3-valved capsules, 6-8 mm long. **Flowering** Spring and summer. **Habitat** Sandy soils in eucalypt forests of the coast and tablelands of NSW and Vic. **Family** Violaceae.

Lobelia dentata

Upright or scrambling annual herb to 40 cm high. **Leaves** Alternate, ovate with deeply-cut margins, mainly around the base of the stem, 15-30 mm long and 3-10 mm wide. **Flowers** Deep-blue, tubular, 10-40 mm long, split to the base with 3 narrow-oblong petals, forming a lower lip and 2 small, curled upper petals, with 5 stamens. Arranged in small terminal racemes of 8-10 flowers. **Fruits** 2-valved ovoid capsules, about 5 mm across. **Flowering** From autumn to spring. **Habitat** Sandy soils in open forests of the coast and tablelands of NSW. **Family** Campanulaceae.

Lobelia gibbosa
Tall Lobelia

Upright annual herb to 65 cm high. **Leaves** Alternate, linear to lanceolate, sometimes slightly toothed, withering early, to 7 cm long and 4 mm wide. **Flowers** Pale to deep bluish-purple with a white streak, tubular, 10-25 mm long, split to the base with 3 narrow oblong petals forming a lower lip, and 2 small, curled upper petals, with 5 stamens. Arranged in one-sided terminal racemes. **Fruits** Swollen, ovoid capsules, 4-10 mm long, with minute, angled seeds. **Flowering** From spring to autumn. **Habitat** Shady sites in dry forests of the coast and tablelands in all states except NT. **Family** Campanulaceae.

Mazus pumilio
Swamp Mazus

Prostrate perennial herb. **Leaves** Ovate to oblong-lanceolate with irregular shallow teeth, slightly hairy, forming a rosette, 8-55 mm long and 2-18 mm wide, on short stalks. **Flowers** Pale-blue to purple or pink with a yellow or white throat, tubular, 7-15 mm long, split to the base with a 2-lobed upper lip and 3-lobed lower lip, with 4 stamens, arranged in a loose raceme of 1-6 flowers on long stalks. **Fruits** 2-valved elliptical capsules, 5-7 mm long. **Flowering** From spring to autumn. **Habitat** Around swamps and boggy sites in coastal NSW, Qld, Vic., SA and Tas. **Family** Scrophulariaceae.

Mazus pumilio

Lobelia dentata

Lobelia gibbosa

Hybanthus monopetalus

Hybanthus vernonii

Utricularia dichotoma

Pandorea doratoxylon

Climbing, multi-stemmed, woody shrub, sometimes regarded as the inland form of *Pandorea pandorana*, with stems to 5 m long. **Leaves** Opposite, divided into 5-9 narrow-lanceolate leaflets, each 2-5 cm long and 2-5 mm wide. **Flowers** Cream with brown-purple markings inside, tubular, 14-25 mm long with a 3-lobed lower lip and 2-lobed upper lip, hairy inside with 4 stamens. They are arranged in short, terminal, leafy racemes. **Fruits** Flat, elliptic, beaked capsules, 7-10 cm long. **Flowering** Winter and spring. **Habitat** Sandy and rocky sites in gorges and sheltered hillsides throughout central areas of all mainland states. **Family** Bignoniaceae.

Pandorea pandorana   Wonga Wonga Vine

Vigorous, woody, climbing or scrambling shrub, with stems to 6 m long. **Leaves** Opposite, 8-16 cm long, divided into 3-9 ovate-oblong to lanceolate leaflets each 2-8 cm long and 2-30 mm wide. **Flowers** Whitish to yellow, often with dark-purple spots or stripes inside, tubular, 10-25 mm long and about 6 mm across with a 3-lobed lower lip and 2-lobed upper lip. They are arranged in loose terminal or axillary clusters. **Fruits** 2-valved oblong capsules, 4-6 cm long and 1-2 cm wide, with winged seeds, 10-15 mm across. **Flowering** Spring. **Habitat** Widespread in wet shady sites on the coast and tablelands of Qld, NSW, Vic., WA and Tas. **Family** Bignoniaceae.

Calostemma purpureum   Garland Lily. Purple Bells

Upright, bulbous, perennial herb to 50 cm high with a thick, fleshy stem. **Leaves** Present at flowering time only. They are long, linear, 4-18 mm wide, arising from the base of the stem. **Flowers** Purple to pink, yellow or rarely white, trumpet-shaped, 5-18 mm long with 6 lobes and 6 yellow stamens, arranged in a terminal cluster of 8-30 flowers on stalks 1-3 cm long. **Fruits** Capsules, 6-14 mm diameter. **Flowering** Summer and autumn. **Habitat** Along the banks of inland rivers of NSW and Vic., and coastal cliffs in SA. **Family** Liliaceae.

Pityrodia teckiana   Native Foxglove

Erect, sticky shrub to 1 m high. **Leaves** Opposite, oblong to lanceolate with deeply-toothed margins, about 2 cm long. **Flowers** Purplish-white to pale-pink, bell-shaped, about 2 cm long with 5 lobes and 4 stamens, arranged in axillary clusters of 1-3 flowers. **Fruits** Dry rounded capsules. **Flowering** Spring and early summer. **Habitat** Granite outcrops in central WA. **Family** Verbenaceae.

Notothixos subaureus   Golden Mistletoe

Parasitic, erect or spreading shrub, to 60 cm across, growing on other mistletoes, covered with yellow downy hairs on young growth. **Leaves** Opposite, elliptical or ovate, yellow-hairy below, 10-50 mm long and 1-3 cm wide. **Flowers** Greenish, covered in golden-yellow down, bell-shaped, 2-8 mm long with 4 stamens, arranged in terminal clusters of 3-7 flowers. **Fruits** Globular berries 5-6 mm diameter. **Flowering** Most of the year. **Habitat** Widespread, parasitic on other epiphytic mistletoes along the coast and adjacent ranges of Qld, NSW and Vic. **Family** Viscaceae.

Lomandra longifolia   Long or Spiny-headed Mat Rush

Upright perennial herb to 1 m high. **Leaves** Narrow-linear to narrow-oblong, flat and glossy, toothed at the tip, 50-100 cm long and 5-14 mm wide. **Flowers** Cream to brown, bell to cup-shaped, stalkless, 3-5 mm long with 6 lobes and 6 stamens, strongly scented, arranged in stiff, stalkless clusters on a tough, flattened stalk, 20-80 cm long. **Fruits** Brown, shiny, ovoid capsules about 6 mm long. **Flowering** Spring and early summer. **Habitat** Widespread on sandy sites, often near watercourses in Qld, NSW, Vic., SA and Tas. **Family** Xanthorrhoeaceae.

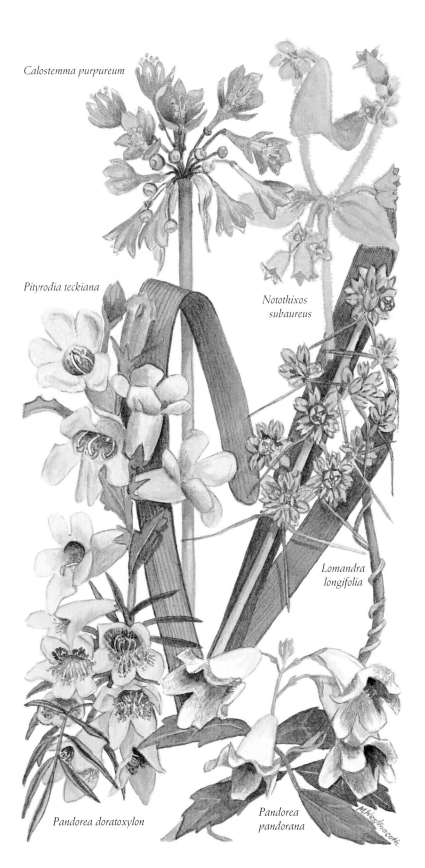

Calostemma purpureum

*Notothixos
subaureus*

Pityrodia teckiana

*Lomandra
longifolia*

Pandorea doratoxylon

*Pandorea
pandorana*

J.M.Westacott.

Epacris impressa
Common Heath

Victoria's floral emblem, an erect, wiry shrub to 2.5 m high. **Leaves** Stalkless, ovate to lanceolate, sharply-pointed, crowded on the stem, 4-16 mm long and 1-6 mm wide. **Flowers** Pink or red, rarely white, bell-shaped, 6-20 mm long and 4-8 mm across, with short stalks, 5 lobes and 5 stamens. They are arranged in leafy spikes. **Flowering** Most of the year. **Habitat** Wet soils in heaths and woodlands of the coast and tablelands of south-eastern NSW, Vic., southeastern SA and Tas. **Family** Epacridaceae.

Rhododendron lochae
Native Rhododendron

Erect or straggling shrub to 2 m high, sometimes growing on trees. **Leaves** Alternate, ovate, thick and glossy to 7 cm long and 4 cm wide. **Flowers** Bright red, bell-shaped, to 5 cm long and 3 cm across with 5 lobes. They are arranged in terminal clusters. **Fruits** Slender capsules 3-4 mm long. **Flowering** Spring and summer. **Habitat** High, mossy rainforest areas and rocky sites in northeastern Qld. **Family** Ericaceae.

Blandfordia grandiflora
Large-flowered Christmas Bell

Upright, tufted, perennial herb, usually 80 cm high, but up to 175 cm. **Leaves** Crowded at the base of the stem, linear, grass-like with rough margins, to 80 cm long and 1-5 mm wide. **Flowers** Red with yellow-tipped lobes, rarely all yellow, bell-shaped, 35-60 mm long and 18-40 mm across with 6 lobes. They are arranged in racemes of 2-20 flowers. **Fruits** Stalked capsules to about 6 cm long. **Flowering** Spring and summer. **Habitat** Damp, sandy or peaty soils of the coast and tablelands from southeastern Qld to central eastern NSW. **Family** Liliaceae.

Blandfordia nobilis
Christmas Bell

Upright, tufted, perennial herb to 80 cm high. **Leaves** Crowded at the base of the stem, grass-like, rigid, to 80 cm long and 1-5 mm wide. **Flowers** Brownish-red with yellow-tipped lobes, bell-shaped, 2-4 cm long and 5-10 mm wide with 6 lobes. They are arranged in racemes of 3-20 flowers. **Fruits** Stalked capsules to about 6 cm long. **Flowering** Spring and summer. **Habitat** Widespread in sandy heaths and coastal swamps in central eastern and southeastern NSW. **Family** Liliaceae.

Blandfordia punicea
Tasmanian Christmas Bell

Upright, tufted, perennial herb to 1 m high. **Leaves** Crowded at the base of the stem, grass-like with slightly curved-back, finely-toothed margins, rigid, to 1 m long and about 1 cm wide. **Flowers** Red outside and yellow inside, rarely all yellow, bell-shaped, 3-5 cm long with 6 lobes. They are arranged in terminal racemes of 5-12 flowers. **Fruits** Stalked capsules. **Flowering** Spring and summer. **Habitat** Wet heaths, moors and hillsides from sea-level to the sub-alps in Tas. **Family** Liliaceae.

Blancoa canescens
Red Bugle. Winter Bell

Upright perennial herb to 50 cm high. **Leaves** Linear, rigid, grass-like, pointed, arising from and sheathing the base of the plant, 7-30 cm long and about 5 mm wide, grey-green with soft hairs and prominent parallel veins. **Flowers** Orange-red to purple-pink, bell-shaped, 3-4 cm long and about 1 cm across with 6 lobes, hairy outside, in short racemes on branching stems. **Flowering** Winter. **Habitat** Sandy heaths and woodlands from Perth to Hill River in WA. **Family** Haemodoraceae.

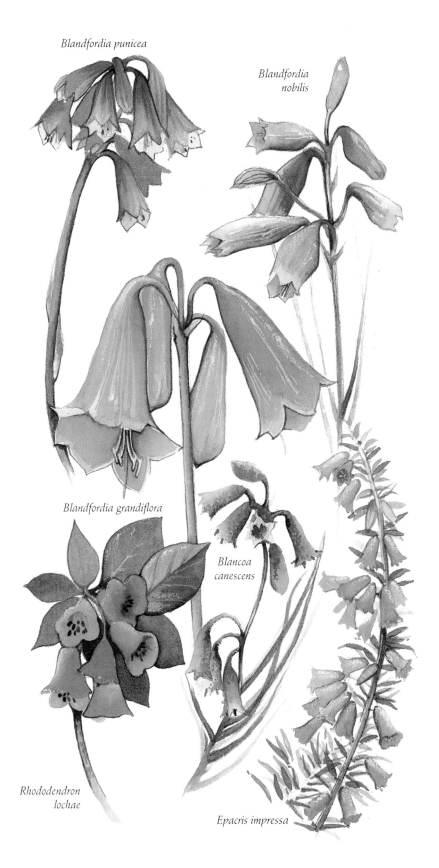

Blandfordia punicea

Blandfordia
nobilis

Blandfordia grandiflora

Blancoa
canescens

Rhododendron
lochae

Epacris impressa

Astroloma conostephioides

Flame Heath

Low, compact, rigid, prickly shrub to 1 m high, with downy branches. **Leaves** Linear to lanceolate, stiff, sharply-pointed with curved-back margins, 10-25 mm long and about 2 mm wide. **Flowers** Bright-red, tubular, 18-25 mm long with downy lobes and overlapping, shiny-red bracts. They are solitary on short axillary stalks. **Fruits** Succulent berries. **Flowering** Winter and spring. **Habitat** Rocky or sandy sites in heaths and open forests in Vic. and SA. **Family** Epacridaceae.

Correa lawrenciana

Mountain Correa

Erect shrub or small tree to 9 m high, with downy branches. **Leaves** Opposite, narrow to broad-elliptic or ovate, hairless or covered with rusty hairs below, smooth, green above, 3-11 cm long and 1-7 cm wide. **Flowers** Greenish-white, bell-shaped, 16-32 mm long and 4-7 mm across, downy outside with 8 long, protruding stamens, arranged in terminal or axillary clusters of 1-7 flowers. **Fruits** Dry capsules to 9 mm long. **Flowering** Mostly in spring. **Habitat** Moist sheltered sites in tall forests and rainforest margins at higher altitudes in southeastern Qld, eastern NSW, Vic. and northeastern Tas. **Family** Rutaceae.

Correa pulchella

Bushy shrub to 60 cm high with downy branches. **Leaves** Opposite, oblong to ovate, dark-green above, paler below, sometimes rough, 1-2 cm long and to 1 cm wide. **Flowers** Pale red to orange, pink or rarely white, bell-shaped, 15-25 mm long, downy outside with 8 protruding stamens, solitary in the leaf axils. **Fruits** Dry capsules. **Flowering** Autumn and winter. **Habitat** Southern coastal areas of SA. **Family** Rutaceae.

Correa reflexa

Native Fuchsia. Common Correa

Variable, erect or spreading, downy shrub, 0.5-1.5 m high. **Leaves** Opposite, lanceolate to heart-shaped, almost stalkless, downy below and sometimes slightly rough above, 2-5 cm long and 6-30 mm wide. **Flowers** Red with greenish-yellow tips or white, cream or green, bell-shaped, 2-4 cm long, downy outside with 8 protruding stamens, terminal, solitary or in threes on short, drooping stalks. **Fruits** Dry capsules 6-9 mm long. **Flowering** Winter and spring. **Habitat** Widespread in dry sclerophyll forests and heaths of southeastern Qld, NSW, Vic., SA, and northeastern Tas. **Family** Rutaceae.

Billardiera longiflora

Purple Apple Berry. Mountain Blue Berry

Twining shrub with wiry stems, climbing to 6 m high. **Leaves** Alternate, narrow-elliptic to narrow-oblong or lanceolate, stiff, dark-green, 10-75 mm long and 3-9 mm wide. **Flowers** Greenish-yellow, sometimes purple-tinted, bell-shaped, 2-3 cm long with 5 slightly protruding stamens, terminal and solitary on long, pendulous stalks. **Fruits** Purple or red, shiny, ovoid berries, 10-25 mm long, with many seeds. **Flowering** Spring and summer. **Habitat** Widespread in damp, shaded sites in sclerophyll forests and woodlands in the northern and southern tablelands of NSW, Vic. and Tas. **Family** Pittosporaceae.

Billardiera scandens

Common Apple Berry

Twining shrub with wiry, hairy stems, climbing to 5 m high. **Leaves** Alternate, narrow-ovate to linear with very short stalks, 1-6 cm long and 2-15 mm wide. **Flowers** Greenish or pale-yellow, rarely orange, purple-tinted when older, bell-shaped, 12-25 mm long with 5 stamens, terminal and usually solitary on long, pendulous stalks. **Fruits** Olive-green, cylindrical berries, 1-4 cm long. **Flowering** Spring and summer. **Habitat** Widespread on sandy soils in heaths and open forests of the coast and tablelands in southeastern Qld, NSW, Vic., SA and Tas. **Family** Pittosporaceae.

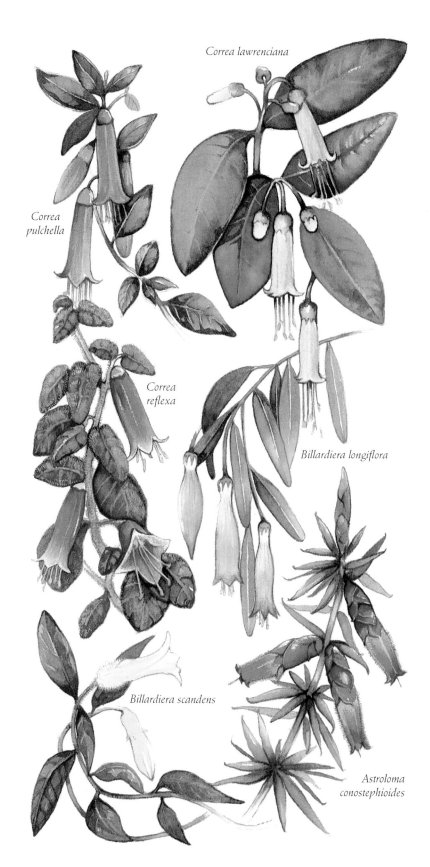

Correa lawrenciana

*Correa
pulchella*

*Correa
reflexa*

Billardiera longiflora

Billardiera scandens

*Astroloma
conostephioides*

Doryanthes palmeri
Spear Lily

Tall, perennial herb with a long flower stem to 5 m high. **Leaves** Fleshy and sword-like, arising from the root system, 1-3 m long and 15 cm or more wide with pointed tips. **Flowers** Red to red-brown, tubular to cup-shaped, about 6-12 cm long with 6 fleshy lobes and 6 conspicuous stamens. They are arranged in a large conical flowerhead to 120 cm long, surrounded by reddish leafy bracts, at the end of a solitary stem with leaves to 30 cm long. **Fruits** Oval, woody, 3-celled capsules, 7-9 cm long, containing winged seeds 15-22 mm long. **Flowering** Late winter and spring. **Habitat** Exposed rocky outcrops in wet sclerophyll forests on the coast and adjacent ranges of northern NSW and southeastern Qld. **Family** Agavaceae.

Sprengelia sprengelioides

Erect or trailing wiry shrub to 1 m high. **Leaves** Ovate to triangular, 3-12 mm long and 1-4 mm wide, sharply-pointed with minutely-toothed margins, stiff, concave, with stem-sheathing bases. **Flowers** White, conical to cup-shaped, about 6 mm long and 6 mm across with 5 lobes and 5 reddish stamens, solitary or in small terminal clusters. **Flowering** Winter and spring. **Habitat** Swampy, sandy heaths from central coastal Qld to the central coast of NSW. **Family** Epacridaceae.

Gentianella diemensis
Mountain Gentian

Erect annual herb to 40 cm high. **Leaves** Opposite, oblong to lanceolate or linear, 1-3 cm long and 2-5 mm wide, mainly around the base of the plant, becoming narrower towards the top of the stem. **Flowers** White or creamy with purplish markings, often with a yellow centre, cup-shaped, 1-2 cm across with 4-5 lobes and 4-5 stamens, either solitary or in small clusters arising from the upper leaf axils. **Fruits** 2-valved cylindrical capsules, 12-17 mm long. **Flowering** Mainly in summer. **Habitat** Grasslands and swampy areas, usually in the alps and sub-alps of NSW, Vic., southeastern SA and Tas. **Family** Gentianaceae.

Guichenotia macrantha
Large-flowered Guichenotia

Erect loosely-branched shrub to 2 m high. **Leaves** Alternate, narrow-oblong, covered with greyish woolly hairs, with a prominent midrib, 3-8 cm long and 4-6 mm wide. **Flowers** Mauve to purplish-brown, cup-shaped with 5 pointed, ribbed lobes and 5 stamens, arranged in racemes of 2-3 flowers on hairy stalks. **Fruits** 5-valved, dry, ovoid capsules, 10-12 mm long. **Flowering** Winter and spring. **Habitat** Sandy or gravelly soils in southwestern WA. **Family** Sterculiaceae.

Sollya heterophylla
Australian Bluebell

Climbing or bushy shrub with twining branches. **Leaves** Alternate, variable, ovate to lanceolate, shiny-green, 3-5 cm long. **Flowers** Blue to white, cup-shaped, 10-18 mm long with 5 overlapping pointed lobes and 5 stamens, arranged in small, pendulous, axillary clusters on long stalks. **Flowering** Spring and summer. **Habitat** Sandy and gravelly soils of the Darling Range and coast of southwestern WA and northeastern Tas. **Family** Pittosporaceae.

Ludwigia peploides
Water Primrose

Creeping or floating perennial herb. **Leaves** Alternate, narrow-ovate to obovate or elliptical, 1-10 cm long and 4-30 mm wide. **Flowers** Yellow, cup-shaped, 1-2 cm long, with 5 overlapping lobes and 10 stamens, solitary on long axillary stalks. **Fruits** Ribbed capsules, 1-3 cm long and 2-4 mm diameter. **Flowering** Summer and autumn. **Habitat** Introduced from South America, a noxious weed, naturalised and widespread on the margins of ponds and creeks in Qld, NSW, Vic and SA. **Family** Onagraceae.

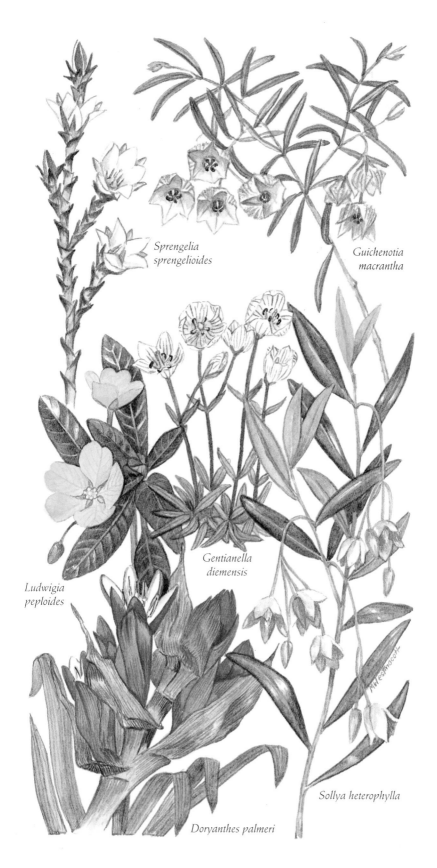

*Sprengelia
sprengelioides*

*Guichenotia
macrantha*

*Gentianella
diemensis*

*Ludwigia
peploides*

Sollya heterophylla

Doryanthes palmeri

Boronia serrulata
Rose Boronia. Sydney Rock Rose

Erect, bushy shrub to 1.5 m high. **Leaves** Opposite, crowded, almost stalkless, rhomboidal with finely-toothed margins, lemon-scented, 7-20 mm long and 5-10 mm wide. **Flowers** Rose-pink, rarely white, aromatic, cup-shaped, 7-12 mm long and about 15 mm across with 4 overlapping lobes and 8 stamens, crowded in leafy terminal clusters of up to 4 flowers, rarely solitary. **Fruits** Capsules, exploding when ripe. **Flowering** Late winter and spring. **Habitat** Sandy soils in moist heaths of the coast and ranges of central eastern NSW. **Family** Rutaceae.

Darwinia collina
Yellow Mountain Bell

Bushy shrub to 1.5 m high. **Leaves** Opposite, crowded, elliptical with finely-toothed margins, leathery, 5-10 mm long and about 5 mm wide. **Flowers** Yellow, cup to bell-shaped, 2-3 cm long with 5 overlapping, rounded lobes and 10 stamens, stalkless, in small, terminal, pendant clusters. **Fruits** Nuts. **Flowering** Spring and summer. **Habitat** Peaty soils above 1000 m in the Stirling Range of WA. **Family** Myrtaceae.

Boronia molloyae
Tall Boronia

Upright, spreading, hairy shrub to 4 m high with erect branches. **Leaves** Opposite, fern-like, divided into 5-15 narrow, flat, dark-green, aromatic leaflets about 1 cm long and 2 mm wide. **Flowers** Pink to red, bell-shaped, about 6 mm long with 4 overlapping, pointed lobes and 8 stamens, solitary and pendant on short axillary stalks. **Fruits** Capsules, exploding when ripe. **Flowering** Spring and summer. **Habitat** Damp, sandy soils in south-western WA. **Family** Rutaceae.

Boronia megastigma
Brown or Sweet-scented Boronia

Erect shrub to 3 m high with wiry branches and a strong aroma. **Leaves** Opposite, divided into usually 3 narrow-linear leaflets, 10-15 mm long. **Flowers** Dark-brown to yellow-green outside and yellow-green inside, very fragrant, cup-shaped, 5-6 mm long and 1 cm across with 4 overlapping lobes and 8 stamens, solitary on short axillary stalks. **Fruits** Capsules, exploding when ripe. **Flowering** Late winter and spring. **Habitat** Swampy sites in forests of southwestern WA. **Family** Rutaceae.

Senna artemisioides
(syn. Cassia artemisioides)
Silver, Woody or Blunt-leaved Cassia

A very variable, dense, rounded shrub, 1-3 m high with many erect, slender branches. There are several sub-species and number of hybrid forms. **Leaves** Variable in size and divided into 1-8 pairs of needle-like to linear, obovate or elliptic leaflets, 7-40 mm long and 2-20 mm wide with stalks 4-60 mm long, with or without silky or woolly hairs, sometimes with curled-under margins, green to grey-green. **Flowers** Yellow, cup-shaped, about 15 mm across and 7-10 mm long with 5 overlapping petals, a cluster of 10 stamens and long, brown anthers. They are arranged in short, dense, axillary racemes of 4-12 flowers. **Fruits** Flat brown pods, 4-8 cm long and 6-10 mm wide, mostly straight. **Flowering** Mainly in winter and spring. **Habitat** Widespread, particularly in dry inland districts on plains and rocky slopes in all mainland states. **Family** Caesalpiniaceae.

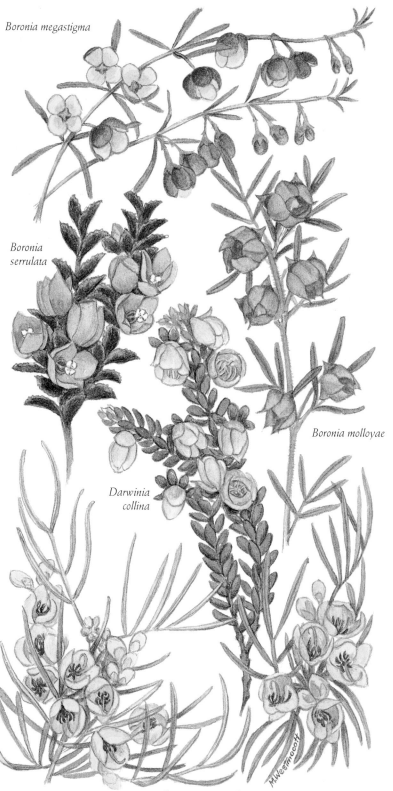

Boronia megastigma

Boronia serrulata

Boronia molloyae

Darwinia collina

Senna artemisioides

M.Westmacott

Convolvulus erubescens
Blushing or Australian Bindweed

Twining or creeping perennial herb, sometimes with hairy stems. **Leaves** Very variable, from ovate-oblong to palm-like, sometimes lobed, sometimes hairy, 1-6 cm long and 3-35 mm wide, with a stalk 5-30 mm long. **Flowers** Pink to white, cup or funnel-shaped with fused, hairy lobes, 7-11 mm long and 15-25 mm across, solitary or 2-3 on axillary stalks. **Fruits** Globular capsules, 4-6 mm across. **Flowering** Mainly in spring and summer. **Habitat** Widespread in grasslands, sclerophyll forests and woodlands in all states. **Family** Convolvulaceae.

Trichodesma zeylanicum
Cattle Bush

Upright or spreading, stiff shrub to 1 m high. **Leaves** Alternate towards the top of the plant, opposite below, narrow-ovate to linear, 3-12 cm long and 7-25 mm wide. **Flowers** Blue, rarely white, cup-shaped, 12-20 mm across with 5 lobes and the stamens in a protruding white column. They are arranged in one-sided racemes on hairy stems. **Flowering** Most of the year. **Habitat** Widespread in dry inland areas, often on sand dunes and rocky hills of Qld, NSW, SA, WA and the NT. **Family** Boraginaceae.

Byblis gigantea
Rainbow Plant

Upright, perennial, insectivorous herb to 60 cm high. **Leaves** Alternate, linear, coiled when young, very narrow, 20-60 cm long, clothed in sticky insect-trapping and digesting glandular hairs. **Flowers** Deep-pink to purple, cup-shaped, 2-4 cm across with 5 rounded lobes. They are solitary on long, glandular, axillary stalks. **Fruits** Flattened capsules. **Flowering** Winter and spring. **Habitat** Sandy soils, waterlogged in winter, dry in summer, in southwestern WA. **Family** Byblidaceae.

Solanum laciniatum
Large-flowered Kangaroo Apple

Erect shrub to 3 m high, often with purplish stems. **Leaves** Variable, deeply-lobed or entire, broad-ovate, 9-38 cm long with lobes 2-13 cm long and 3-20 mm wide. Entire leaves are lanceolate, 5-20 cm long and 1-4 cm wide. **Flowers** Blue to violet, broadly cup-shaped, 3-5 cm across with 5 shallowly-notched lobes and 5 yellow stamens, arranged on stalks in loose clusters of up to 11 flowers. **Fruits** Orange-yellow ovoid berries 15-20 mm across. **Flowering** Spring and summer. **Habitat** Sandy sites along creeks and roadsides of the coast and tablelands in northern and southern NSW, Vic., southeastern SA, southwestern WA and northeastern Tas. **Family** Solanaceae.

Duboisia hopwoodii
Pituri

Compact, erect shrub to 4 m high and 3 m across, with many branches and corky bark. **Leaves** Alternate, narrow-elliptic or lanceolate with a pointed or hooked tip, thick, 2-15 cm long and 2-13 mm wide. **Flowers** White with purple stripes, funnel or cup-shaped, 7-15 mm long and 4-8 mm across with 5 rounded lobes and 4 stamens, arranged in terminal clusters. **Fruits** Purple-black succulent berries, 3-6 mm across with brown seeds 2-3 mm long. **Flowering** Mainly from late winter to summer. **Habitat** Widespread in arid regions on red sand soils, mallee and woodlands, in all mainland states except Vic. **Family** Solanaceae.

Boronia polygalifolia
Milkwort. Dwarf Boronia

Erect or prostrate shrub with trailing stems to 80 cm long. **Leaves** Opposite, narrow-ovate to narrow-lanceolate or spathulate, fleshy with pointed tips, 5-30 mm long and 1-6 mm wide. **Flowers** White to deep pink, cup-shaped, 8-15 mm across with 4 overlapping lobes and 8 stamens, usually solitary in the leaf axils on stalks 2-6 mm long. **Fruits** Hairless capsules, exploding when ripe. **Flowering** Mainly in spring and summer. **Habitat** Open forests and heaths on damp, rocky slopes of the coast and tablelands, in central eastern and southeastern Qld and eastern NSW. **Family** Rutaceae.

Boronia polygalifolia

Byblis gigantea

Trichodesma zeylanicum

Solanum laciniatum

Duboisia hopwoodii

Convolvulus erubescens

Solanum elegans (syn. S. amblymerum)   Spiny Kangaroo Apple

Erect shrub to 1 m high with prickly branches, stems and leaves. **Leaves** Alternate, hairy, linear to narrow-lanceolate, toothed or entire, 4-6 cm long and 5-15 mm wide with sharp spines 1-8 mm long, protruding from the upper midrib. **Flowers** Purple, broadly cup-shaped, 2-3 cm across with 5 papery lobes and 5 yellow stamens, arranged in terminal clusters of 1-6 flowers. **Fruits** Red berries, 10-15 mm across. **Flowering** Late winter and summer. **Habitat** Moist rocky sites in sclerophyll forests of the coast and tablelands in southeastern Qld and NSW. **Family** Solanaceae.

Solanum aviculare   Kangaroo Apple

Erect shrub to 4 m high with purplish-green, angular stems. **Leaves** Alternate, variable, lobed or entire, deep-green. Lobed leaves are obovate to elliptic, 15-30 cm long with lobes 1-10 cm long and 5-20 mm wide. Entire leaves are lanceolate to narrow-elliptic, 8-25 cm long and 10-35 mm wide. **Flowers** Blue or violet, broadly cup-shaped, 25-40 mm across with 5 papery lobes and 5 yellow stamens, arranged in loose axillary clusters of up to 11 flowers on stalks to 15 cm long. **Fruits** Orange-red ovoid berries, 10-15 mm diameter. **Flowering** Summer. **Habitat** Wet forests and rainforest margins, and disturbed sites in Qld, NSW, Vic., southern SA and southwestern WA. **Family** Solanaceae.

Solanum brownii   Violet Nightshade

Erect, hairy shrub to 2 m high, sometimes with prickly branches. **Leaves** Alternate, narrow-ovate to elliptic, dark-green above, with grey or rusty hairs below, usually with sharp spines protruding from the upper midrib, 4-12 cm long and 1-3 cm wide. **Flowers** Pale-blue to purple, broadly cup-shaped, 15-40 mm across with 5 papery lobes and 5 yellow stamens, arranged in terminal clusters of 1-10 flowers on stalks 1-2 cm long. **Fruits** Yellow, greenish or white berries, 15-20 mm diameter. **Flowering** Mainly in winter and spring. **Habitat** Sclerophyll forests and woodlands, disturbed rainforests and pastures on the coast and ranges of southeastern Qld, NSW and southeastern Vic. **Family** Solanaceae.

Solanum campanulatum

Prickly, perennial shrub to 1 m high with hairy stems and spines on the stalks and branches. **Leaves** Alternate, ovate to elliptical, 8-13 cm long and 5-10 cm across, with shallow lobes and sharp spines and matted hair on both surfaces. **Flowers** Purple, broadly cup-shaped, 1-3 cm across with 5 papery lobes and 5 yellow stamens, with a spiny calyx, arranged in loose clusters of 4-10 flowers on spiny stalks to 3 cm long. **Fruits** Pale-green to yellow, globular berries, drying black, 20-25 mm diameter. **Flowering** Year round. **Habitat** Widespread in sclerophyll forests along the coast and ranges of southeastern Qld, northeastern and central eastern NSW. **Family** Solanaceae.

Solanum sturtianum   Thargomindah Nightshade

Erect shrub to 3 m high, covered with silvery-green or greyish downy hairs, sometimes with spines on the branches. **Leaves** Alternate, lanceolate, covered in downy hairs, silvery below, pale-green above, 3-6 cm long and 5-15 mm wide. **Flowers** Purple, broadly cup-shaped, 3-4 cm across with 5 papery lobes and 5 yellow stamens, arranged in terminal clusters of 1-12 flowers. **Fruits** Brittle, yellow to brownish-black berries, 10-15 mm diameter. **Flowering** Year round. **Habitat** Dry inland areas in all mainland states except Vic. **Family** Solanaceae.

Solanum vescum   Gunyang

Erect or spreading shrub to 2 m high with hairy new growth and no prickles. **Leaves** Alternate, often deeply-lobed. Lobed leaves are broad-ovate with lobes 5-10 cm long and 8-12 mm wide with stalks to 3 cm long. Entire leaves are linear to lanceolate, with a short stalk or stalkless, 5-15 cm long and 5-13 mm wide. **Flowers** Pale-mauve to purple, broadly cup-shaped, 3-4 cm across with 5 papery lobes and 5 yellow stamens, arranged in loose clusters on stalks to 5 cm long. **Fruits** Greenish berries, 20-25 mm diameter. **Flowering** Winter and spring. **Habitat** Sandy areas along the coast of southeastern Qld, NSW, eastern Vic. and Tas. **Family** Solanaceae.

Solanum elegans

Solanum brownii

Solanum sturtianum

Solanum aviculare

Solanum vescum

Solanum campanulatum

Cryptandra tomentosa
Prickly Cryptandra

Low, tangled, rigid shrub to 1.5 m high. **Leaves** Alternate, clumped on small branches, needle-like, 2-6 mm long. **Flowers** White, rarely red, cup-shaped, 2-3 mm long with 5 pointed lobes and 5 stamens, woolly outside, arranged in small terminal clusters. **Fruits** Capsules. **Flowering** Late winter and spring. **Habitat** Sandy sites in open forests, heaths and mallee scrub, inland in Vic., SA and northeastern Tas. **Family** Rhamnaceae.

Micromyrtus ciliata
Heath Myrtle

Erect or spreading shrub to 1.2 m high. **Leaves** Opposite, decussate, crowded, linear to lanceolate or obovate, 1-4 mm long and 1-3 mm wide, thick and fleshy with glandular dots, and minute hairs on the margins. **Flowers** White or pink, tubular to cup-shaped, 1-4 mm long and 3-4 mm across with 5 rounded lobes. They are solitary, almost stalkless, and crowded in the upper leaf axils. **Fruits** Reddish angular nuts. **Flowering** Spring and early summer. **Habitat** Widespread in sandy or rocky areas in heaths, open forests and scrubs, inland and along the coast and tablelands of central and southeastern NSW, Vic and southeastern SA. **Family** Myrtaceae.

Nymphoides geminata
Fringed Waterlily. Entire Marshwort

Aquatic. perennial or annual herb with floating or semi-erect leaves and upright flower stems to 1 m long. **Leaves** Heart-shaped to orbicular, 1-8 cm long and 2-8 cm wide, with stalks to 20 cm long. **Flowers** Yellow, tubular, 2-3 cm across with 5 (sometimes 6) fringed, spreading lobes. They are arranged in pairs or clusters on long stalks along the flowering stem. **Fruits** Brown, turning black, capsules, about 1 cm long. **Flowering** From spring to autumn. **Habitat** Ponds, swamps and rivers in water up to 1 m deep along the coast and adjacent plateaus of Qld, NSW southeastern Vic and Kangaroo Is. in SA. **Family** Menyanthaceae.

Nymphoides crenata
Wavy Marshwort

Aquatic perennial herb with floating leaves and upright flower stems. **Leaves** Arise from the base of the pant, they are orbicular to kidney-shaped with rounded teeth along the margins, 5-20 cm long and up to 15 cm wide, dotted with small glands below. **Flowers** Yellow to orange, tubular, 2-5 cm across with 4-6 (usually 5) fringed, spreading lobes. They are arranged in pairs or clusters on stalks 3-8 cm long along the flower stem. **Fruits** Ellipsoid capsules, 4-12 mm long. **Flowering** From spring to autumn. **Habitat** Slow-flowing water up to 1.5 m deep, widespread inland in all mainland states. **Family** Menyanthaceae.

Valorise exaltata
Yellow Marsh Flower

Erect, tufted, aquatic perennial herb with stems to 1.5 m long. **Leaves** Ovate to broad-ovate or kidney-shaped, sometimes with wavy margins, floating, 4-15 cm long and 2-10 cm wide, with stalks 13-40 cm long. **Flowers** Yellow, tubular, 16-40 mm across, with 4-6 (usually 5) spreading lobes, bearded inside and fringed at the base, arranged in loose clusters on stout, erect, flowering stems. **Fruits** 4-valved capsules, 6-12 mm long, with seeds to 3 mm long. **Flowering** From spring to early autumn. **Habitat** Peaty swamps and ephemeral pools in water to 60 cm deep, on the coast and ranges of southeastern Qld, NSW, Vic. and northeastern Tas. **Family** Menyanthaceae.

Mirabilis jalapa
Four O'clock. Marvel of Peru

Upright, bushy, perennial herb to 1 m high with stout stems. **Leaves** Opposite, ovate to triangular or lanceolate, 4-9 cm long and 1-5 cm wide, on stalks 1-4 cm long. **Flowers** White, yellow or red, tubular to funnel-shaped, 2-6 cm long and up to 35 mm across with 5 rounded, spreading lobes and 5-6 protruding stamens, opening in cloudy weather or late afternoon, with a strong perfume at night. They are arranged in small terminal clusters of 3-7 flowers, but only the central one develops. **Fruits** Sub-globular black nuts about 7 mm long. **Flowering** Autumn. **Habitat** Introduced from South America, naturalised along the coast of Qld, southeastern and central eastern NSW, Vic. and the Southern Lofty region of SA. **Family** Nyctaginaceae.

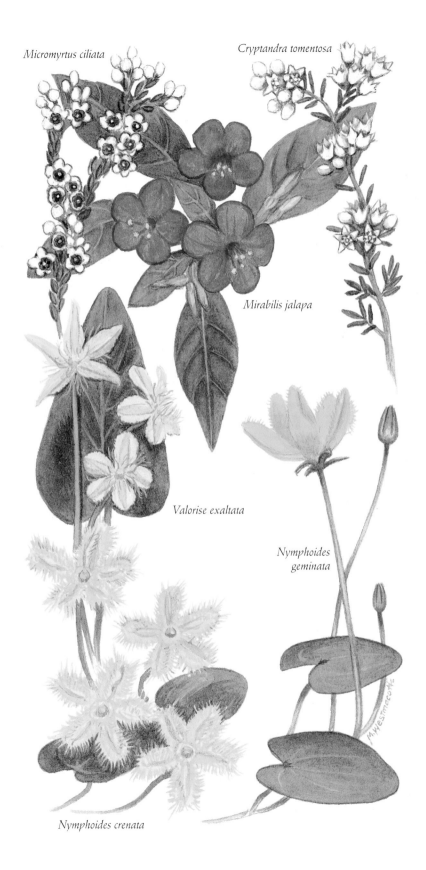

Micromyrtus ciliata

Cryptandra tomentosa

Mirabilis jalapa

Valorise exaltata

Nymphoides geminata

Nymphoides crenata

Goodenia affinis
Silver or Cushion Goodenia
Tufted, decumbent or ascending, usually perennial herb with stems to 20 cm long. **Leaves** Oblong to ovate, concave, 2-4 cm long and 5-15 mm wide, forming a rosette around the base of the plant, covered with white woolly hairs. **Flowers** Yellow, tubular, woolly outside, 12-20 mm long with 5 widely-spreading, rounded lobes, split to the base, 2 arching over the pollen cup, with 5 stamens. They are solitary on axillary stalks 1-2 cm long, or in racemes to 5 cm long. **Fruits** Ovoid capsules, 10-15 mm long, containing smooth, brown, elliptic seeds, 2 mm long. **Flowering** Late winter and spring. **Habitat** Widespread in mallee woodlands from southeastern SA to the central west coast of WA. **Family** Goodeniaceae.

Goodenia geniculata
Bent Goodenia
Low or ascending, tufted, perennial herb to 50 cm high, covered with short hairs when young. **Leaves** Linear to narrow-elliptic, arising from the base of the plant, sometimes with toothed margins, slightly hairy, 2-10 cm long and 2-20 mm wide. **Flowers** Yellow with brown stripes, tubular, hairy outside, about 25 mm across with 5 widely-spreading, rounded lobes, 2 overlapping the pollen cup, with 5 stamens. They are arranged in leafy racemes to 40 cm long. **Fruits** Ovoid capsules to 1 cm long. **Flowering** Most of the year. **Habitat** Widespread in sclerophyll forests and open woodlands of western Qld, the northwestern plains of NSW, Vic., southern SA, western and southwestern Tas. **Family** Goodeniaceae.

Goodenia hederacea
Ivy or Forest Goodenia
Prostrate or ascending perennial herb to 80 cm high. **Leaves** Oval, elliptic or egg-shaped, with entire or toothed margins, 1-12 cm long and 3-25 mm wide with cottony hairs below. **Flowers** Yellow, tubular, cottony outside, with soft hairs inside, 8-15 mm long and to 2 cm across with 5 notched, widely-spreading lobes, 2 overlapping the pollen cup. They are arranged in leafy racemes to 80 cm long, on slender stalks to 5 cm long. **Fruits** Globular to ovoid capsules, 5-9 mm long, with pale-yellow seeds about 3 mm long. **Flowering** Spring and summer. **Habitat** Widespread in forests, woodlands and grasslands of the coast and ranges of southeastern Qld, NSW, central and eastern Vic. **Family** Goodeniaceae.

Goodenia ovata
Hop Goodenia
Weak, erect or scrambling shrub to 2.5 m high, often with sticky young shoots. **Leaves** Alternate, ovate or elliptical, with finely-toothed margins, 2-10 cm long and 1-6 cm wide. **Flowers** Yellow, tubular, 1-2 cm long and about 2 cm across, with 5 widely-spreading, rounded lobes, 2 arching over the pollen cup, arranged in groups of 3-6 in the leaf axils on stalks to 4 cm long. **Fruits** Narrow cylindrical capsules, 8-15 mm long. **Flowering** Mainly in spring and summer. **Habitat** Widespread in forests and woodlands, and along coastal cliffs and headlands in southeastern Qld, eastern NSW, Vic., SA and Tas. **Family** Goodeniaceae.

Goodenia stelligera
Spiked Goodenia
Erect, tufted, perennial herb to 60 cm high. **Leaves** In a rosette around the base of the plant, linear to narrow-lanceolate, broader towards the tip, sometimes irregularly toothed, thick and glossy, 5-25 cm long and 1-12 mm wide. **Flowers** Yellow, tubular, 13-16 mm long, hairy outside with 5 widely-spreading, rounded lobes, 2 overlapping the pollen cup, solitary or in clusters of 2-3. **Fruits** Ovoid to oblong capsules, 5-9 mm long with brown seeds about 2 mm long. **Flowering** Spring and summer. **Habitat** Widespread in wet sites of eastern Qld, NSW and southeastern Vic. **Family** Goodeniaceae.

Lechenaultia biloba
Blue Lechenaultia
Low, straggling or ascending perennial herb or shrub to 1 m high. **Leaves** Alternate, needle-like, crowded, 6-15 mm long, soft and slightly fleshy. **Flowers** Blue, shortly tubular, 14-20 mm long and 2-3 cm across with 5 wedge-shaped lobes, 2 smaller than the others, delicate, hairy inside, in small clusters in the upper leaf axils. **Fruits** Flattish capsules 15-35 mm long and about 3 mm wide. **Flowering** Late winter and spring. **Habitat** Widespread on gravelly hillsides in southwestern WA. **Family** Goodeniaceae.

Goodenia stelligera

Goodenia geniculata

Goodenia ovata

Lechenaultia biloba

Goodenia affinis

Goodenia hederacea

Leucopogon parviflorus
Coastal Bearded Heath

Erect shrub or small tree, low on exposed sites, 1-5 m high. **Leaves** Elliptic to lanceolate, broadest towards the tip, stiff, paler below, 1-3 cm long and 2-8 mm wide. **Flowers** White, tubular with 5 hairy, rounded, spreading lobes, 2-4 mm long and 4-6 mm across, arranged in short, terminal, leafy spikes, 10-35 mm long, in tight clusters of 7-13 flowers. **Fruits** Round, whitish, fleshy drupes, 3-5 mm across. **Flowering** Most of the year. **Habitat** Widespread and common on sand dunes and sandy coastal heaths in all states except the NT. **Family** Epacridaceae.

Leucopogon virgatus
Common Bearded Heath

Low shrub to 60 cm high with wiry, downy, red-brown branches. **Leaves** Narrow-lanceolate to ovate, concave, thick, sometimes toothed, 3-22 mm long and 1-4 mm wide, on very short stems. **Flowers** White to pinkish, tubular with 5 spreading, hairy lobes, 1-3 mm long and 3-6 mm across, arranged in short terminal spikes, 5-10 mm long, of 4-7 flowers. **Fruits** Oblong drupes 3-5 mm long. **Flowering** From late winter to early summer. **Habitat** Sandy soils in heaths and open forests of the coast and tablelands of Qld, NSW, Vic., SA and Tas. **Family** Epacridaceae.

Pandorea jasminoides
Bower Vine

Vigorous, twining, climbing, woody shrub. **Leaves** Opposite, sometimes in whorls of 3, divided into 4-7 lanceolate leaflets each 45-60 mm long and 15-30 mm wide. **Flowers** White to pink with a maroon, hairy throat, tubular, 4-6 cm long and 4-6 cm across with 5 spreading lobes and 4 stamens, arranged in terminal clusters. **Fruits** Oblong to ovoid, beaked, woody pods, 4-6 cm long and 1-2 cm across, with winged seeds 10-15 mm across. **Flowering** Spring and summer. **Habitat** Wet areas in forests and rainforests of southeastern Qld and northern NSW. **Family** Bignoniaceae.

Myoporum acuminatum
(syn. M. montanum)
Water Bush. Native Myrtle

Erect, bushy shrub or small tree, to 13 m high, with fissured bark. **Leaves** Alternate, narrow-elliptic to lanceolate, 3-14 cm long and 2-38 mm wide, sometimes with toothed margins. **Flowers** White with purple spots, tubular with a hairy throat, 5-8 mm across, with 5 spreading, rounded lobes and 4 protruding stamens. They are arranged in axillary clusters of 1-8 flowers. **Fruits** Ovoid, purple to purplish-black drupes, 4-8 mm long, smooth or wrinkled. **Flowering** Winter and spring. **Habitat** Sclerophyll forests, mallee and rainforests on the coast and inland in all mainland states. **Family** Myoporaceae.

Acrotriche prostrata
Trailing Ground Berry

Prostrate trailing shrub with stems to 2 m long. **Leaves** Opposite, broad-lanceolate, rigid, 1-2 cm long and to 1 cm wide, hairy above. **Flowers** Greenish, tubular, 4-8 mm long with 5 spreading lobes tipped with long hairs. They are arranged in dense axillary clusters on very short stalks. **Flowering** Winter. **Habitat** Cool sites in forests in Vic. **Family** Epacridaceae.

Pentachondra pumila
Carpet or Cushion Heath

Prostrate perennial woody herb to 12 cm high, forming dense mats to 1 m across. **Leaves** Opposite, crowded and stalkless, leathery, oblong to elliptic, concave with fine hairs on the margins, 3-6 mm long and about 3 mm wide. **Flowers** White, occasionally red, tubular, 3-4 mm long and 4-6 mm across with 5 spreading, bearded lobes. They are solitary and stalkless at the ends of short branches. **Fruits** Crimson berries 5-8 mm diameter. **Flowering** Summer. **Habitat** Widespread above the tree line in heaths, tall herbfields and tussock grasslands in alpine areas of southern NSW, Vic. and Tas. **Family** Epacridaceae.

Myoporum acuminatum

Leucopogon parviflorus

Leucopogon virgatus

Pandorea jasminoides

Pentachondra pumila

Acrotriche prostrata

DEVAUX

Brunoniella australis
Blue Trumpet

Slender perennial herb, prostrate or erect, to 30 cm high, sparsely covered with hairs. **Leaves** Variable, broad-ovate to obovate or lanceolate, hairy, 12-80 mm long and 5-30 mm wide. **Flowers** Blue, rarely white, tubular, 6-12 mm long and 10-15 mm across, hairy with 5 spreading, rounded lobes, solitary or in small clusters on short stalks in the upper leaf axils. **Fruits** Ovoid capsules, 10-16 mm long. **Flowering** From spring to winter. **Habitat** Moist areas in forests and woodlands, along the coast, ranges and western slopes and plains of Qld, NSW and northern NT. **Family** Acanthaceae.

Mentha australis
River Mint

Slender, sprawling or erect, soft, aromatic, perennial herb to 1 m high with quadrangular stems, often purplish. **Leaves** Opposite, narrow-ovate with toothed margins, hairy, 1-6 cm long and 5-20 mm wide. **Flowers** White to pink, tubular, hairy outside, 5-7 mm long with 4 unequal, spreading lobes and 4 protruding stamens, arranged in dense clusters of 3-12 flowers in the leaf axils. **Fruits** Small ovoid nuts. **Flowering** From late summer to early spring. **Habitat** Widespread in clay soils, particularly near waterways, inland in all states except WA. **Family** Lamiaceae.

Diplarrena moraea
Butterfly Flag. White Iris

Upright, tufted, perennial herb to 1 m high. **Leaves** Narrow-linear, 10-70 cm long and 5-10 mm wide, dark-green, arising from and sheathing the base of the plant. **Flowers** White with yellow and sometimes purple markings, shortly tubular, 5-7 cm across with 3 large, spreading, outer lobes and 3 smaller inner lobes, often yellow-tinged. They are honey-scented and arranged in a terminal cluster of 2-6 flowers on a stem 20-100 cm long. **Fruits** Cylindrical capsules, 20-25 mm long and about 7 mm wide with brown seeds about 3 mm across. **Flowering** Spring. **Habitat** Widespread in heaths and sclerophyll forests, often on granite, along the south coast and tablelands of NSW, southern Vic. and Tas. **Family** Iridaceae.

Chamelaucium uncinatum
Geraldton Wax Flower

Brittle spreading shrub to 5 m high. **Leaves** Opposite, thick, linear, 1-4 cm long and about 1 mm wide with a hook at the tip. **Flowers** White, red, purple or pink, funnel-shaped, about 6 mm long and 15-25 mm across with 5 spreading, rounded lobes surrounding the tube which is fringed with 10 stamens and has a central protruding style. They are arranged in terminal clusters of 2-4 flowers. **Fruits** Hard capsules. **Flowering** Spring and summer. **Habitat** Common on limestone hills of the east coast of WA. **Family** Myrtaceae.

Mimulus repens
Creeping Monkey Flower

Prostrate or creeping annual or perennial herb, to 20 cm high, forming dense mats, occasionally aquatic. **Leaves** Opposite, crowded and stalkless, fleshy, concave, elliptic to ovate, 1-10 mm long and 1-7 mm wide. **Flowers** Blue, violet or pink with yellow and white markings inside, tubular, 5-12 mm long with 5 unequal, spreading lobes and 4 stamens, solitary in the upper leaf axils. **Fruits** Globular capsules about 5 mm across. **Flowering** From spring to autumn. **Habitat** Brackish marshes and swamps in all states except WA and the NT. **Family** Scrophulariaceae.

Scaevola striata
Blue Fan Flower. Royal Robe

Spreading, ascending or prostrate perennial herb to 20 cm high with hairy stems. **Leaves** Obovate to wedge-shaped, toothed, hairy, 1-15 cm long and 3-25 mm wide. **Flowers** Purple to blue with red stripes, bearded inside, tubular, 13-27 mm long with 5 spreading, rounded lobes and 5 yellow stamens, solitary or in racemes to 30 cm long, in the leaf axils. **Fruits** Ovoid to oblong drupes, 5-7 mm long. **Flowering** Spring and summer. **Habitat** Forested areas in southwestern WA. **Family** Goodeniaceae.

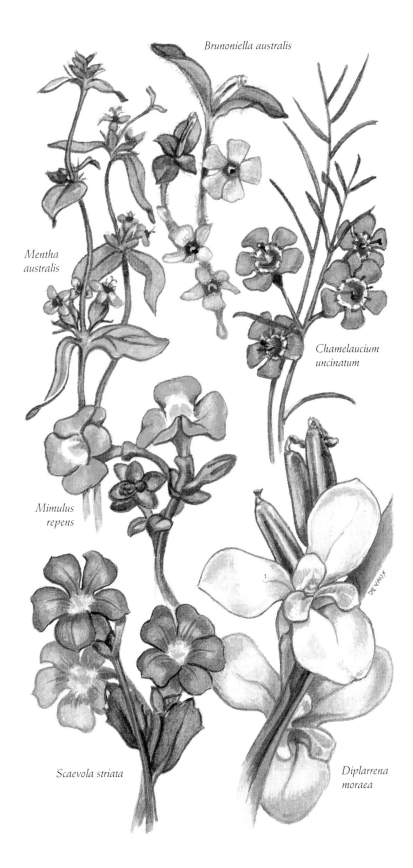

Brunoniella australis

Mentha australis

Chamelaucium uncinatum

Mimulus repens

Scaevola striata

Diplarrena moraea

Dampiera purpurea

Erect, many-stemmed, perrenial, weak, hairy shrub to 1.5 m high. **Leaves** Obovate to elliptic, thick, covered with grey or brown hairs below, 1-6 cm long and 5-42 mm wide, sometimes with toothed margins. **Flowers** Purple to blue with a light-yellow throat, tubular, 12-15 mm long and about 25 mm across with 5 notched, spreading lobes covered with dark woolly hairs outside, solitary or in small terminal or axillary clusters. **Fruits** Ribbed, hairy nuts, 4-5 mm long. **Flowering** Spring and summer. **Habitat** Widespread on sandy soils in heaths and open forests of the coast and ranges of southeastern Qld, eastern NSW and southeastern Vic. **Family** Goodeniaceae.

Dampiera stricta   Blue Dampiera

Small, erect or straggling, perennial shrub to 60 cm high with angular stems. **Leaves** Elliptical to lanceolate or linear, sometimes with a few coarse teeth, semi-succulent, stalkless, bunched towards the top of the stem, 10-65 mm long and 2-25 mm wide. **Flowers** Blue, rarely white, with a pale-yellow throat, tubular, 10-14 mm long with 5 spreading, unequal, notched lobes, rusty-hairy outside, solitary or in axillary clusters of up to 3 flowers. **Fruits** Ribbed, rusty-hairy nuts, 4-5 mm long. **Flowering** Spring and summer. **Habitat** Sandy soils in heaths and open forests of the coast and ranges in southeastern Qld, eastern NSW, central and eastern Vic. southwestern WA and northeastern Tas. **Family** Goodeniaceae.

Coopernookia barbata

Erect, hairy shrub to 60 cm high. **Leaves** Spirally arranged, oblong-linear to lanceolate with curled-under margins, sometimes toothed, sticky hairy, 1-5 cm long and 1-8 mm wide. **Flowers** Blue to pinkish-purple, tubular, 11-15 mm long with 5 spreading, unequal, notched lobes, hairy outside, 2 arching over the pollen cup, arranged in terminal leafy racemes. **Fruits** Cylindrical to ovoid capsules, 5-7 mm long with seeds 4-5 mm long. **Flowering** Year round. **Habitat** Widespread in dry sclerophyll forests of the coastal mountains of NSW and southeastern Vic. **Family** Goodeniaceae.

Scaevola aemula   Fairy or Common Fan Flower

Upright or sprawling perennial herb with ascending stems, covered in coarse yellow hairs, to 50 cm high. **Leaves** Alternate, obovate to elliptical, coarsely-toothed, 1-9 cm long and 4-35 mm wide. **Flowers** Pale-blue to mauve, often with a yellow throat, tubular, hairy, stalkless, 15-25 mm long and 12-30 mm across with 5 lobes spreading to one side of the flower, fan-like, arranged in leafy, terminal spikes to 24 cm long. **Fruits** Ovoid, downy, wrinkled drupes, 3-5 mm long. **Flowering** Mainly in late winter, spring and early autumn. **Habitat** Widespread in dry sclerophyll forests on sandy soils, sand dunes and cliffs along the coast of Qld, NSW, Vic., SA, southern WA and northeastern Tas. **Family** Goodeniaceae.

Scaevola ramosissima   Purple or Hairy Fan Flower

Straggling, hairy, rough, annual herb to 40 cm high. **Leaves** Linear to lanceolate, sometimes toothed, hairy, thick, 2-10 cm long and 2-10 mm wide. **Flowers** Blue to mauve, tubular, 15-25 mm long, split down one side with 5 fan-like spreading lobes, hairy inside, solitary or in leafy clusters to 30 cm long. **Fruits** Ellipsoid, hairy, wrinkled drupes, 5-7 mm long. **Flowering** Most of the year. **Habitat** Widespread on sandy soils in heaths and dry sclerophyll forests of eastern Qld, eastern NSW and southeastern Vic. **Family** Goodeniaceae.

Scaevola spinescens   Currant Bush. Spiny Fan Flower

Rigid, erect, hairy shrub to 2 m high, sometimes with terminal and lateral spines. **Leaves** Clustered along the branches, obovate to linear, thick, greyish-green, 9-36 mm long and 1-6 mm wide. **Flowers** White to pale-yellow, sometimes with thin purple stripes, tubular, slit to the base on one side, 9-20 mm long with 5 lobes spreading to one side of the flower, fan-like, bearded inside, solitary on slender axillary stalks. **Fruits** Black or purple, ovoid, fleshy drupes, 5-8 mm across. **Flowering** Most of the year. **Habitat** Widespread in drier areas on hillsides or stony sites inland in all mainland states. **Family** Goodeniaceae.

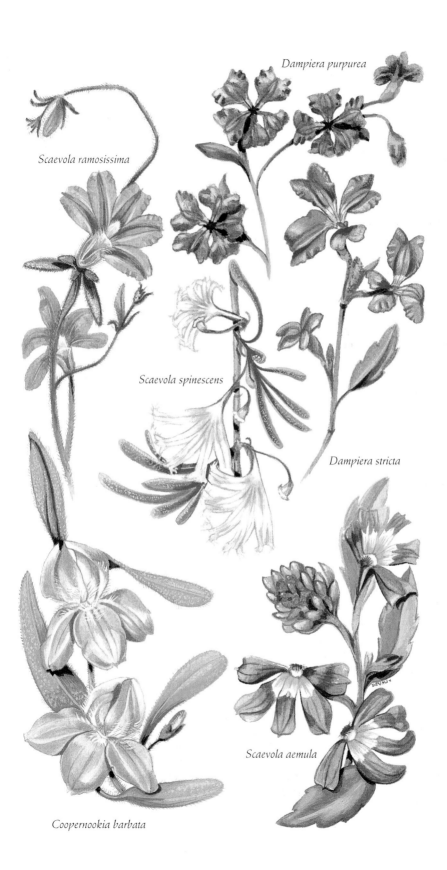

Dampiera purpurea

Scaevola ramosissima

Scaevola spinescens

Dampiera stricta

Scaevola aemula

Coopernookia barbata

Dendrophthoe vitellina   **Long-flower Mistletoe**

Parasitic shrub attached to tree branches. **Leaves** Mostly alternate, elliptic to narrow-lanceolate, leathery with a prominent midrib, 4-16 cm long and 6-30 mm wide. **Flowers** Yellow to scarlet, tubular, 25-50 mm long with 5 thin, scarlet, linear, spreading lobes, and 5 long, protruding, yellow stamens. They are arranged in dense axillary racemes of 5-20 flowers. **Fruits** Yellow to red ovoid berries, 10-15 mm long. **Flowering** Spring and summer. **Habitat** Parasitic on many hosts in dry sclerophyll forests, especially Rough-barked Apple, on the coast and ranges of Qld, NSW and southeastern Vic. **Family** Loranthaceae.

Lysiana subfalcata   **Northern Mistletoe**

Parasitic shrub attached to tree branches, with stems to 50 cm long. **Leaves** Opposite, narrow-spathulate, 2-12 cm long and 4-20 mm wide, with distinct veins. **Flowers** Red to yellow, tubular, 25-50 mm long with 6 narrow, spreading lobes and 6 long, protruding, red stamens. They are arranged in axillary pairs. **Fruits** Pale, ellipsoid or pear-shaped berries, 8-14 mm long. **Flowering** Most of the year. **Habitat** Parasitic on many hosts, especially *Acacias*, inland in Qld, NSW, SA, WA and the NT. **Family** Loranthaceae.

Amylotheca dictyophleba

Parasitic shrub attached to tree branches, with stems to 1 m long. **Leaves** Mainly opposite, thick and leathery, shiny above and dull below, lanceolate to elliptic or broad-obovate, 6-13 cm long and 2-6 cm wide with a stalk 2-8 mm long. **Flowers** Orange to red with greenish tips, tubular, 3-4 cm long with 4-6 narrow, spreading lobes when opened. They are arranged in axillary clusters of 1-6 pairs or triads. **Fruits** Red or purple, globular berries, 5-12 mm across. **Flowering** Summer and autumn. **Habitat** Parasitic on rainforest trees along the coast and ranges of Qld and northern and central eastern NSW. **Family** Loranthaceae.

Amyema congener

Parasitic shrub attached to tree branches, with stems to 40 cm long. **Leaves** Mainly opposite, lanceolate to ovate or obovate, leathery, light-green, 3-12 cm long and 1-5 cm wide. **Flowers** Green to yellow with a reddish base, tubular, 16-35 mm long with 4-6 narrow, spreading lobes and long, scarlet, protruding stamens. They are arranged in dense heads of 3-5 groups of 3 flowers. **Fruits** Green, globular berries about 8 mm long. **Flowering** Most of the year. **Habitat** Common parasite on *Casuarinas*, *Acacias* and *Geijera parviflora* on the coast and ranges of Qld and eastern NSW. **Family** Loranthaceae.

Amyema maidenii

Parasitic shrub attached to tree branches, usually with hairy stems. **Leaves** Opposite, ovate to broad-spathulate, leathery, grey, 2-6 cm long and 5-23 mm wide. **Flowers** Grey outside and green inside, tubular, 14-30 mm long with 5 narrow, spreading lobes. They are arranged in a head of 2 stalkless triads. **Fruits** Ovoid berries about 8 mm long. **Flowering** Year round. **Habitat** Parasitic on *Acacias,* inland in all mainland states except Vic. **Family** Loranthaceae.

Muellerina eucalyptoides   **Creeping Mistletoe**

Parasitic shrub attached to tree branches, with stems to 2 m long. **Leaves** Opposite, narrow-oblong to lanceolate, light-green, 5-25 cm long and 7-28 mm wide. **Flowers** Reddish-green, tubular, 30-45 mm long with 5 narrow, spreading lobes. They are solitary or arranged in 3-4 pairs of 3-flowered clusters or single flowers along short branches. **Fruits** Yellow, pear-shaped berries 8-15 mm long. **Flowering** Mainly in summer. **Habitat** Usually parasitic on *Eucalyptus* trees, widespread along the coast and tablelands of southeastern Qld, NSW, Vic. and southeastern SA. **Family** Loranthaceae.

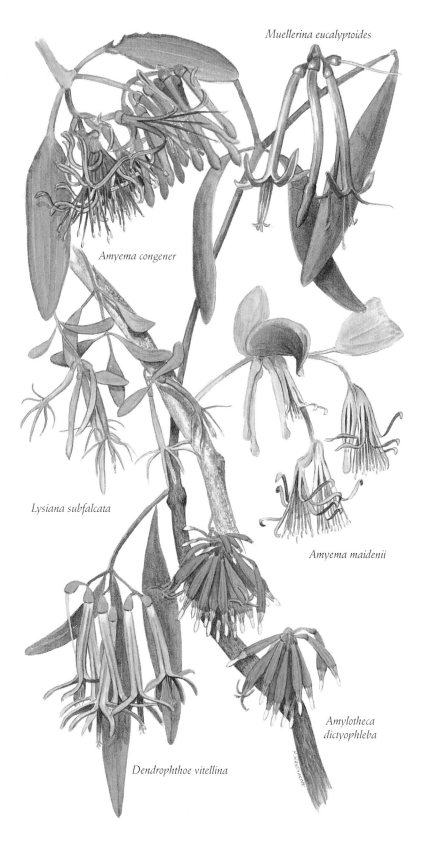

Muellerina eucalyptoides

Amyema congener

Lysiana subfalcata

Amyema maidenii

Dendrophthoe vitellina

Amylotheca dictyophleba

Doryanthes excelsa
Gymea, Flame or Giant Lily

Tall perennial herb with a long, straight flowering stem to 5 m high. **Leaves** arise from the base of the plant. They are fibrous and sword-like, 1-2.5 m long and 15 cm or more wide with a pointed tip and distinct spines on the margins, arising from the root system. **Flowers** Pinkish-red or rarely white, tubular, 10-16 cm long with 6 spreading lobes and large quantities of nectar. They are arranged in a large, globular cluster to 70 cm across at the end of a solitary flowering stem with leaves to 30 cm long. **Fruits** Woody, ovoid, 3-celled capsules, 7-10 cm long, containing many flat, winged seeds, 15-23 mm long. **Flowering** Spring. **Habitat** Coast and adjacent ranges of northern and central NSW. **Family** Agavaceae.

Crinum flaccidum
Darling, Macquarie or Murray Lily

Erect, bulbous, clumping, perennial herb to 70 cm high with a large, fleshy, flattened stem. **Leaves** Broad-linear, usually with rough margins, 30-80 cm long and 1-5 cm wide, arising from the base of the stem. **Flowers** White or yellow, heavily scented, tubular, 4-12 cm long with 6 pointed, spreading lobes and 6 protruding stamens, terminal in clusters of 5-16 on long stalks at the end of the stout, fleshy stem. **Flowering** From summer to early autumn. **Habitat** Moist inland areas along river banks and floodplains in all mainland states. **Family** Liliaceae.

Crinum pedunculatum
Swamp Lily

Erect, bulbous, perennial herb to 80 cm high with a large, fleshy, flattened stem. **Leaves** Linear, thick and broad, 50-200 cm long and 5-15 cm wide, arising from the base of the stem. **Flowers** White to pale-mauve, scented, tubular, 4-10 cm long and 4-8 cm across with 6 long, narrow, curled, spreading lobes, and 6 long, protruding stamens. They are arranged in clusters of 10-40 flowers at the end of a stout, fleshy stem. **Fruits** Rounded capsules 2-5 cm across with a prominent beak. **Flowering** Spring and summer. **Habitat** Wet and swampy sites including rainforests of coastal Qld, NSW and NT. **Family** Liliaceae.

Anthocercis littorea
Coast Rayflower. Yellow Tailflower

Erect or spreading shrub to 3 m high with many slender branches. **Leaves** Oblong to obovate, broader towards the tip, slightly toothed, fleshy with a prominent mid-vein, 18-65 mm long and 4-31 mm wide. **Flowers** Yellow with purple-red streaks in the throat, strongly perfumed, tubular, 6-10 mm long and 14-35 mm across with 5 narrow, spreading lobes and 4 stamens, solitary or in small axillary clusters on thin stalks. **Fruits** 2-valved globular capsules, 9-19 mm long with seeds 2 mm long. **Flowering** Autumn, winter and spring. **Habitat** Sandy coastal sites in western and southern WA. **Family** Solanaceae.

Correa alba
White Correa

Compact, erect shrub to 1.5 m high, with rusty hairs on the young stems. **Leaves** Opposite, ovate to orbicular, thick, dull-green above and covered with whitish or rusty hairs below, 15-40 mm long and 6-30 mm wide. **Flowers** White or pinkish, waxy, tubular at the base, 11-15 mm long and 2 cm across with 4 spreading, pointed lobes and 8 stamens, arranged in terminal clusters of 1-4 flowers. **Fruits** Green capsules 5-7 mm long. **Flowering** Mainly in winter. **Habitat** Sandy and rocky coastal sites in NSW, Vic., southeastern SA and Tas. **Family** Rutaceae.

Eichhornia crassipes
Water Hyacinth

Aquatic, perennial, floating herb to 1 m high. **Leaves** arise from the base of the plant, swollen and spongy, elliptical or orbicular, to 25 cm diameter, on stalks 5-75 cm long. **Flowers** Lilac with yellow and blue markings on the upper lobe, tubular, about 15 mm long and 3-6 cm across with 6 spreading, pointed lobes and 6 stamens, arranged in short axillary racemes of about 8 flowers, to 20 cm long. **Fruits** 3-celled capsules. **Flowering** Most of the year. **Habitat** Introduced from South America, a noxious weed in Australia, found in nutrient-rich streams and ponds in all mainland states. **Family** Pontederiaceae.

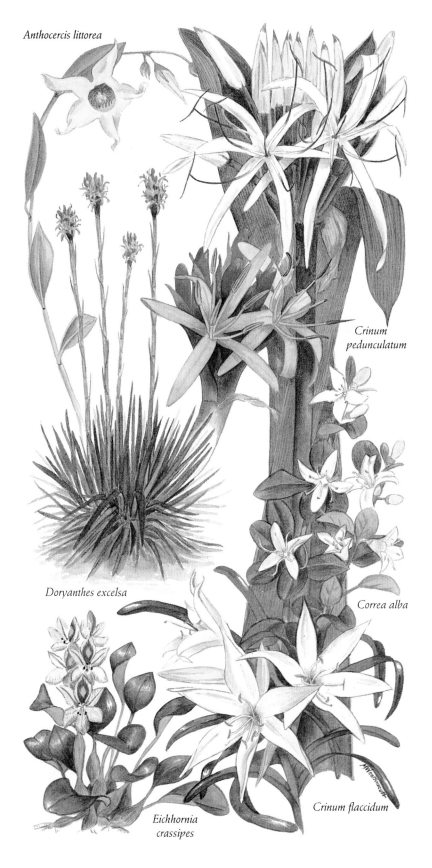

Anthocercis littorea

Crinum pedunculatum

Doryanthes excelsa

Correa alba

Eichhornia crassipes

Crinum flaccidum

Isotoma axillaris
Showy Isotome

Upright or prostrate, branching, perennial herb to 50 cm high, with purplish stems and an acrid, poisonous sap. **Leaves** Alternate, 15-150 mm long, deeply cut into narrow-linear, pale-green segments, often with toothed margins. **Flowers** Pale-blue to mauve, rarely white or pink, tubular, 15-35 mm long with 5 spreading, pointed lobes. They are terminal and solitary on slender axillary stalks, 3-17 cm long. **Fruits** Cylindrical capsules 7-18 mm long. **Flowering** From spring to autumn. **Habitat** Rocky cliffs in shallow, sandy soils, often around waterholes, widespread in southeastern Qld, NSW and Vic. **Family** Campanulaceae.

Isotoma fluviatilis
Swamp Isotome

Prostrate or creeping perennial herb with stems to 20 cm long, forming dense mats. **Leaves** Alternate, spathulate, ovate to orbicular, sometimes slightly toothed, 4-15 mm long and 2-8 mm wide. **Flowers** Pale-blue to purple and white, tubular, 4-10 mm long with 5 spreading, pointed lobes, hairy inside with 5 fused yellow stamens, solitary and terminal on slender, axillary stalks, 5-30 mm long. **Fruits** Cylindrical to ovoid capsules, 3-6 mm long. **Flowering** Spring and summer. **Habitat** Widespread in moist areas of the coast and ranges in southeastern Qld, eastern NSW, Vic., southeastern SA and Tas. **Family** Campanulaceae.

Isotoma petraea
Rock Isotome

Upright perennial herb to 40 cm high, often forming dense clumps. **Leaves** Alternate, ovate to lanceolate or elliptic, coarsely-toothed or lobed, 15-75 mm long and 5-55 mm wide. **Flowers** White or pale-green to lilac, tubular, 5-13 mm long with 5 spreading, pointed lobes and 5 yellow stamens, solitary on axillary stalks, 8-25 cm long. **Fruits** Ovoid capsules, 10-22 mm long. **Flowering** Most of the year. **Habitat** Rocky outcrops and hillsides, inland in all mainland states except Vic. **Family** Campanulaceae.

Wahlenbergia gloriosa
Royal Bluebell

Slender, perennial herb, with erect stems, 6-40 cm high. **Leaves** Confined to the lower stem, opposite or alternate, obovate, becoming lanceolate further up the stem, usually with slightly toothed margins, 4-35 mm long and 1-15 mm wide. **Flowers** Blue to deep purple, tubular, 2-3 cm across with 5 delicate, spreading, pointed lobes and 5 stamens. They are solitary on thin stalks, 4-25 cm long. **Fruits** Dry capsules, 7-12 mm long. **Flowering** Summer. **Habitat** Alpine and subalpine woodlands, alpine herbfields and grasslands of the southern tablelands of NSW and Vic. **Family** Campanulaceae.

Wahlenbergia stricta
Tall Bluebell

Upright, tufted, perennial herb to 90 cm high. **Leaves** Confined to the lower stem, stalkless, opposite, becoming alternate up the stem, narrow-oblong to obovate at the base, becoming linear higher up, with wavy or slightly toothed margins, hairy, 5-70 mm long and 1-13 mm wide. **Flowers** Blue, sometimes white, tubular, 4-11 mm long and 2-4 cm across with 5 delicate, spreading, pointed lobes and 5-6 stamens. They are solitary on long thin stalks. **Fruits** Globular to ovoid capsules, 3-10 mm long. **Flowering** Year round. **Habitat** Widespread in a wide variety of sites and plant communities in all states except the NT. **Family** Campanulaceae.

Wahlenbergia multicaulis (syn. W. tadgellii)
Tadgell's Bluebell

Upright, tufted, perennial herb to 75 cm high with long, wiry stems. **Leaves** Alternate or nearly opposite on the lower stem, obovate or lanceolate on the lower stem, linear higher up, with a prominent midrib, stalkless, 4-80 mm long and 1-6 mm wide, sometimes with small teeth on the margins. **Flowers** Blue, shortly tubular, 1-3 cm across with 4-6 (usually 5) delicate, spreading, pointed lobes and 5 stamens. They are solitary on long thin stalks. **Fruits** Dry, broadly-ovoid, ribbed capsules, 4-13 mm long. **Flowering** Year round. **Habitat** Forests, woodlands and grasslands along the coast and ranges of central and central eastern NSW, Vic., southeastern SA, southwestern WA and Tas. **Family** Campanulaceae.

Wahlenbergia stricta

Wahlenbergia multicaulis

Wahlenbergia gloriosa

Isotoma petraea

Isotoma fluviatilis

Isotoma axillaris

Alyxia buxifolia
Sea Box
Compact shrub to 2.5 m high, although usually less than 50 cm high. **Leaves** Opposite or in whorls of 3, obovate to broad-elliptical, 1-4 cm long and 5-25 mm wide, tough, glossy dark-green above and paler below. **Flowers** White with an orange tube, fragrant, 6-9 mm long with 5 spreading, overlapping lobes, twisted sideways. They are arranged in axillary or terminal clusters of about 8 flowers on short stalks. **Fruits** Fleshy red berries, 6-10 mm diameter. **Flowering** From spring to autumn. **Habitat** Exposed cliffs and sand dunes in southeastern NSW, Vic., SA, western and southern WA and Tas. **Family** Apocynaceae.

Monotoca scoparia
Prickly Broom Heath
Compact, erect, rigid shrub, 30-250 cm high. **Leaves** Alternate, oblong to linear or elliptical, flat or with curved-back margins, sharply pointed, thick, 6-15 mm long and 1-3 mm wide, whitish below with fine longitudinal veins. **Flowers** White, tubular, 1-3 mm long and 2-3 mm across with 5 spreading, pointed lobes and 5 reddish-yellow stamens. They are arranged in axillary clusters of 2-6 on very short stalks. **Fruits** Ovoid to oblong, greenish-yellow to orange drupes, 2-3 mm long. **Flowering** Autumn. **Habitat** Sandy soils in heaths and open forests of the coast and tablelands in Qld, NSW, Vic., SA and Tas. **Family** Epacridaceae.

Derwentia perfoliata *(syn. Parahebe perfoliata)*
Digger's Speedwell
Upright, wiry, perennial herb to 1.5 m high. **Leaves** Opposite, stem-clasping, ovate to lanceolate with a few prominent teeth, blue-green, 15-80 cm long and 14-50 mm wide. **Flowers** Blue to lilac, streaked with purple, shortly-tubular, 15-20 mm across, with 4 spreading, pointed lobes and 2 protruding yellow stamens. They are arranged in loose racemes, 10-45 cm long with 25-75 flowers. **Fruits** Ovoid to oblong compressed capsules. **Flowering** Mainly in summer. **Habitat** Moist rocky sites along the coastal tablelands and subalpine areas of NSW and Vic. **Family** Scrophulariaceae.

Stackhousia pulvinaris
Alpine Stackhousia
Prostrate, perennial herb, to 10 cm high, forming dense mats. **Leaves** Alternate, fleshy, crowded, linear to oblong, 5-10 mm long and 1-2 mm wide. **Flowers** White to yellow, sweetly perfumed, shortly-tubular, 3-6 mm long and 8-15 mm across with 5 spreading lobes and 5 stamens. They are solitary in the upper leaf axils. **Fruits** Small nuts. **Flowering** Summer. **Habitat** Wet sites in alpine herbfields and subalpine grasslands of southern NSW, south-eastern Vic. and Tas. **Family** Stackhousiaceae.

Stackhousia viminea
Slender Stackhousia
Upright perennial herb to 70 cm high with longitudinally-ribbed stems. **Leaves** Alternate, larger around the base of the plant, linear to obovate or narrow-elliptic, 4-40 mm long and 2-8 mm wide. **Flowers** Greenish-yellow, tubular, 2-6 mm long with 5 spreading, pointed lobes and 5 stamens, arranged in clusters of 1-5 flowers along the flowering stems. **Fruits** Dry, rough and globular, 2-5 mm across. **Flowering** From spring to autumn. **Habitat** Widespread in forests and woodlands of the coast and ranges in all states except WA and the NT. **Family** Stackhousiaceae.

Calectasia cyanea *(syn. C. intermedia)*
Blue Tinsel Lily
Upright, tufted, perennial herb to 1.5 m high, with wiry stems. **Leaves** Crowded, narrow-linear, pointed, 3-15 mm long and about 1 mm wide. **Flowers** Blue to deep-purple, tubular, 7-12 mm long with 6 widely-spreading, narrow, pointed lobes 10-15 mm long, and 6 protruding yellow stamens. They are stalkless, solitary and terminal on short branchlets. **Flowering** Spring. **Habitat** Moist, well-drained heaths in Vic., SA. and WA. **Family** Xanthorrhoeaceae.

Derwentia perfoliata

Alyxia buxifolia

Calectasia cyanea

Monotoca scoparia

Stackhousia viminea

Stackhousia pulvinaris

Astrotricha ledifolia
Common Star Hair

Erect, slender, hairy shrub to 1.5 m high. **Leaves** Alternate, linear to oblong-linear with curved-back margins, 2-5 cm long and 1-8 mm wide, dark-green and rough above, covered with light-brown hairs below. **Flowers** White to creamy-green, tubular, about 2 mm long and 5 mm across with 5 spreading, pointed lobes and 4-5 stamens. They are arranged in terminal clusters, 7-15 cm long. **Fruits** Dry capsules, separating into segments. **Flowering** Spring and early summer. **Habitat** Hilly sites, heaths, open forests and woodlands above 750 m, in the central and southern tablelands of southeastern Qld, NSW and eastern Vic. **Family** Araliaceae.

Pimelea axiflora
Bootlace Bush. Tough Rice Flower

Erect shrub to 3 m high with long, slender branches. **Leaves** Opposite, linear to elliptic, dark-green above and paler below, 5-80 mm long and 2-10 mm wide, usually with curled-back margins. **Flowers** Creamy-white, small, tubular, 3-9 mm long and 2-3 mm across, with 4 spreading, pointed lobes and 2 yellow stamens. They are almost stalkless in axillary clusters of 2-10 flowers. **Fruits** Nuts about 4 mm long with a succulent red covering. **Flowering** Spring. **Habitat** Moist forested slopes and gullies, heaths and woodlands along the coast and ranges of southeastern NSW, Vic. and northeastern Tas. **Family** Thymelaeaceae.

Pimelea flava
Yellow Rice Flower

Erect, well-branched shrub to 1 m high with hairy young stems and poisonous flowers. **Leaves** Opposite, crowded, stalkless, concave, narrow-elliptic to elliptic, 2-10 mm long and 1-10 mm wide. **Flowers** Yellow or white, small, tubular, about 5 mm across with 4 spreading, pointed lobes and 2 yellow stamens. They are massed in small terminal flowerheads of 9-28 flowers, 10-15 mm across, on stalks 1-10 mm long. **Fruits** Single-seeded ovoid drupes. **Flowering** Spring. **Habitat** Sandy soils in mallee communities, also in damp heaths and cool open forests of the coast and ranges, in central and southwestern NSW, Vic., SA, the south coast of WA and northeastern Tas. **Family** Thymelaeaceae.

Pimelea microcephala

Erect, open shrub to 4 m high, toxic to stock. **Leaves** Opposite, narrow-elliptic to linear, dull pale-green, 7-45 mm long and 1-5 mm wide. **Flowers** Greenish-yellow, small, tubular, males are 3-8 mm long and females 2-4 mm long. They have 4 spreading lobes, silky-hairy, arranged in terminal heads of 13-100 male flowers or 7-12 females. **Fruits** Green or red, ovoid, succulent drupes, 4-6 mm long. **Flowering** Most of the year. **Habitat** Widespread in sandy soils in open forests and mallee, inland in all mainland states. **Family** Thymelaeaceae.

Samolus repens
Creeping Brookweed

Erect or prostrate perennial herb with wrinkled or warty stems to 60 cm long, forming dense tufts. **Leaves** Alternate or arising from the roots, fleshy, obovate to spathulate, 1-5 cm long and 2-12 mm wide, with small linear to lanceolate stem leaves. **Flowers** White, tinged with pink, with a short, broad tube, 6-10 mm long and about 1 cm across with 5 spreading, pointed lobes and 5 stamens surrounding a greenish centre. They are arranged in terminal or sub-terminal, leafy racemes. **Fruits** 5-valved capsules. **Flowering** From spring to autumn. **Habitat** Grows near brackish water and salty swamps along the coast of southeastern Qld, NSW, Vic., SA, WA and Tas. **Family** Primulaceae.

Smilax australis
Sarsparilla. Bush Lawyer

Vigorous, climbing shrub with slender, wiry and prickly branches to 8 m long. **Leaves** Alternate, broadly-ovate to elliptical, leathery with prominent veins, 5-15 cm long and 2-10 cm wide on twisted stalks, 5-15 mm long. **Flowers** Greenish-white to cream, small, broadly-tubular, 1-2 mm long and 3-4 mm across with 6 spreading, pointed lobes and 6 long, protruding stamens in the male flowers. They are arranged in dense axillary clusters of 15-20 flowers. **Fruits** Red to black ovoid berries, 5-12 mm diameter, **Flowering** Mainly in spring and summer. **Habitat** Widespread in moist coastal forests, rainforests, woodlands and heaths, often in dense thickets, along the coast and ranges of Qld, NSW, southeastern Vic., northern WA and northern NT. **Family** Smilacaceae.

Pimelea microcephala

Pimelea flava

Pimelea axiflora

Smilax australis

Astrotricha ledifolia

Samolus repens

Epacris lanuginosa
Woolly Heath

Erect shrub to 2 m high with slender, woolly branches. **Leaves** Spirally arranged, crowded, narrow-lanceolate, sharply-pointed, 6-12 mm long and about 2 mm wide. **Flowers** White, stalkless, tubular, about 8 mm long and 7 mm across with 5 spreading, triangular lobes. They are arranged in dense, terminal, cylindrical, leafy spikes. **Fruits** Capsules about 2 mm long. **Flowering** Spring and summer. **Habitat** Wet heaths and swampy areas up to about 1000 m in Vic. and Tas. **Family** Epacridaceae.

Epacris microphylla
Coral Heath

Erect, wiry shrub to 1.8 m high with hairy branches. **Leaves** Crowded, almost stalkless, heart-shaped, sharply-pointed and concave, 2-6 mm long and 1-6 mm wide. **Flowers** White or pinkish, stalkless, tubular, 2-3 mm long and 3-7 mm across with 5 spreading, broadly-triangular lobes. They are arranged in dense, terminal, cylindrical, leafy spikes, 4-6 mm across. **Fruits** Capsules about 2 mm long. **Flowering** Mainly in winter and spring. **Habitat** Widespread in damp soils in swampy heaths and open forests of the coast and tablelands to about 1900 m in Qld, NSW, Vic. and Tas. **Family** Epacridaceae.

Epacris paludosa
Alpine or Swamp Heath

Erect, bushy shrub to 1.5 m high, with bristly branchlets. **Leaves** Crowded, almost stalkless, lanceolate to ovate, sharply-pointed with tiny teeth on the margins, thick, 5-12 mm long and 2-3 mm wide. **Flowers** White, stalkless, tubular, 5-7 mm long and 5-8 mm across with 5 spreading, pointed lobes. They are arranged in dense terminal clusters, 6-8 mm across. **Fruits** Capsules about 3 mm long. **Flowering** Mainly in spring and summer. **Habitat** Marshy, sub-alpine heaths and montane forests to about 1700 m, in central eastern and southeastern NSW, southeastern Vic. and northeastern Tas. **Family** Epacridaceae.

Brachyloma daphnoides
Daphne Heath

Erect shrub to 2 m high with bristly branchlets. **Leaves** Alternate, lanceolate to ovate, terminally crowded, 4-20 mm long and 2-4 mm wide, glossy-green above and paler below. **Flowers** Honey-scented, cream, tubular, 3-6 mm long and 3-6 mm across with 5 spreading, pointed lobes, produced in the leaf axils along the stems. **Fruits** Green to yellow-brown, flattened, ridged, globular drupes, about 4 mm across. **Flowering** Spring and early summer. **Habitat** Common on poor soils in dry, rocky or sandy areas in open forests of Qld, NSW, Vic. and SA. **Family** Epacridaceae.

Leucopogon australis
Spike Bearded Heath

Erect shrub to 1.5 m high with pale, ridged branches. **Leaves** Lanceolate, dark-green, 2-7 cm long and 2-8 mm wide, with distinct longitudinal veins above. **Flowers** White, tubular, about 3 mm long and 4 mm across with 5 spreading, hairy lobes, arranged in axillary spikes. **Fruits** Yellow to white fleshy drupes. **Flowering** Spring. **Habitat** Near-coastal heaths and low forests of Vic., SA, southwestern WA and Tas. **Family** Epacridaceae.

Myoporum floribundum
Slender Myoporum

Erect, slender shrub to 4 m high, with a sour smell. **Leaves** Alternate to opposite, linear, 2-11 cm long and 1-3 mm wide, dark-green, hanging from the branches. **Flowers** White to pale mauve, shortly tubular, 2-3 mm long and 5-8 mm across with 5 spreading, pointed lobes and long, protruding stamens. They are arranged in axillary clusters of 6-8 flowers along the upper side of the branches. **Fruits** Oblong, compressed drupes, 2-3 mm long. **Flowering** From winter to summer. **Habitat** Rocky slopes in sclerophyll forests of the coast and ranges in central eastern NSW and southeastern Vic. **Family** Myoporaceae.

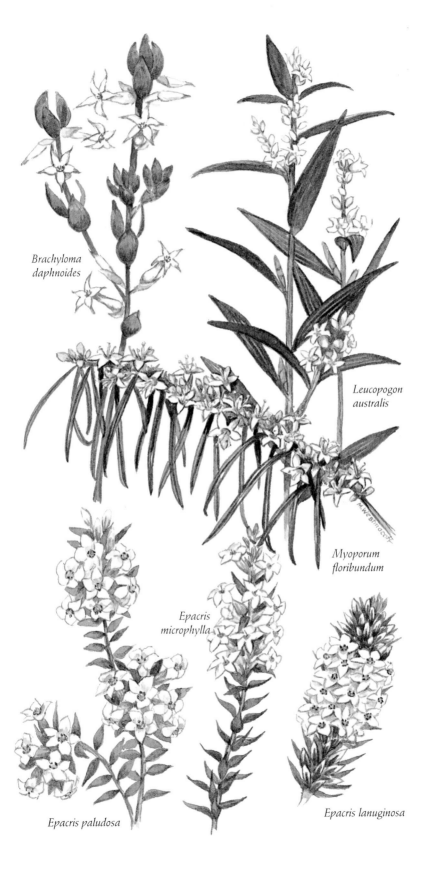

Brachyloma
daphnoides

Leucopogon
australis

Myoporum
floribundum

Epacris
microphylla

Epacris paludosa

Epacris lanuginosa

m.westmacott

Leucopogon lanceolatus
Lance Bearded Heath
Bushy shrub to 3 m high with slender branches. **Leaves** Ovate to lanceolate, paler below, with distinct longitudinal veins, 6-40 mm long and 1-6 mm wide. **Flowers** White, tubular, 1-2 mm long and 3-4 mm across with 5 spreading, pointed, hairy lobes, arranged in terminal slender spikes, 3-40 mm long, of 2-16 flowers. **Fruits** Red, fleshy berries, 2-4 mm diameter. **Flowering** Throughout the year. **Habitat** Most soil types in eucalypt forests and woodlands of the coast and tablelands, from central eastern Qld to NSW, Vic., southeastern SA and Tas. **Family** Epacridaceae.

Sprengelia incarnata
Pink Swamp Heath
Erect shrub to 2 m high with stiff stems. **Leaves** Ovate to lanceolate, 5-20 mm long and 2-6 mm wide, stiff, concave, stem-sheathing at their bases, tapering to a sharp point. **Flowers** Pink, shortly-tubular, 6-16 mm across with 5 narrow, pointed, spreading lobes and protruding red stamens. They are arranged in dense, terminal, leafy, pyramidal heads of 3-20 flowers. **Fruits** Capsules about 2 mm across. **Flowering** Winter and spring. **Habitat** Wet sandy soils in heaths and swamps of the coast and tablelands from northeastern NSW to Vic., southeastern SA and Tas. **Family** Epacridaceae.

Lythrum salicaria
Purple Loosestrife
Upright perennial herb to 1 m high, with angular stems, often covered with soft hairs. **Leaves** Opposite or whorled, stalkless, narrow-ovate, 17-70 mm long and 3-12 mm wide, pointed, slightly stem-clasping. **Flowers** Pink to purple, tubular, ribbed, 7-12 mm long with 5-6 spreading, pointed lobes and 12 stamens. They are arranged in leafy, axillary clusters. **Fruits** Oblong, 2-valved capsules. **Flowering** Spring and summer. **Habitat** Widespread on the edges of swamps and watercourses in eastern Qld, NSW, Vic., southeastern SA and northeastern Tas. **Family** Lythraceae.

Derwentia derwentiana
Derwent Speedwell
(syn. Parahebe derwentiana)
Upright perennial herb to 1.5 m high with a woody base. **Leaves** Opposite, stalkless, narrow-ovate to broad-lanceolate with toothed margins, 5-20 cm long and 12-45 mm wide. **Flowers** Blue, pale-lilac or white, small, 5-9 mm across, shortly-tubular with 4 pointed, spreading lobes and 2 long, protruding stamens, arranged in dense axillary racemes, 8-25 cm long, with 40-100 flowers. **Fruits** Compressed obovate capsules, 3-6 mm long. **Flowering** Mainly in summer. **Habitat** Moist, shaded forest sites to alpine herbfields, along the coast and tablelands of southeastern Qld, NSW, Vic., southeastern SA and Tas. **Family** Scrophulariaceae.

Stackhousia monogyna
Creamy Candles. Creamy Stackhousia
Upright perennial herb to 80 cm high. **Leaves** Alternate, linear to lanceolate, thin, 1-3 cm long and 2-4 mm wide. **Flowers** White to yellow, tubular, 5-8 mm long with 5 spreading lobes and 5 stamens, arranged in a one-sided, cylindrical, terminal spike. **Flowering** Spring and summer. **Habitat** Widespread in heaths and grasslands, sclerophyll forests and woodlands in all states except the NT. **Family** Stackhousiaceae.

Stackhousia spathulata
Semi-prostrate to erect perennial herb, to 50 cm high. **Leaves** Alternate, fleshy, spathulate to obovate, sometimes pointed, 15-30 mm long and 3-15 mm wide. **Flowers** White to cream, tubular, 6-8 mm long with 5 spreading lobes and 5 stamens, arranged in dense, cylindrical, terminal spikes. **Flowering** From early spring to summer. **Habitat** Widespread in heaths, dry sclerophyll forests and sand dunes, often near lagoons along the coast of Qld, NSW, Vic., SA and Tas. **Family** Stackhousiaceae.

Leucopogon
lanceolatus

Sprengelia
incarnata

Lythrum salicaria

Stackhousia
monogyna

Derwentia derwentiana

Stackhousia
spathulata

Epacris obtusifolia
Blunt-leaf Heath

Erect shrub to 2 m high with hairy branchlets. **Leaves** Stem-clasping, elliptical to lanceolate with blunt tips, 5-12 mm long and 2-3 mm wide. **Flowers** Creamy-white, honey-scented, stalkless, tubular with 5 curled-back petals, 5-15 mm long and 3-8 mm across. They are arranged in leafy, terminal, one-sided racemes, 4-10 mm across. **Fruits** Capsules, 3-4 mm long. **Flowering** Mainly from winter to summer. **Habitat** Marshy grounds in heaths and woodlands of the coast and tablelands to about 1000 m, from southeastern Qld through NSW to southwestern Vic. and Tas. **Family** Epacridaceae.

Anigozanthos flavidus
Tall or Evergreen Kangaroo Paw

Erect, hairy, perennial herb to 3 m high. **Leaves** Strap-like, arising from and sheathing the base of the plant, 35-100 cm long and 5-20 mm wide. **Flowers** Yellow-green, occasionally red or orange, tubular, 30-45 mm long with 6 pointed lobes on one side, densely-hairy outside, arranged in terminal racemes. **Fruits** 3-celled capsules. **Flowering** Spring and summer. **Habitat** Swampy sites in forests, woodlands and heaths in southwestern WA. **Family** Haemodoraceae.

Anigozanthos manglesii   Red and Green or Mangles' Kangaroo Paw

Floral emblem of WA, an erect, hairy, perennial herb to 1 m high. **Leaves** Strap-like, arising from and sheathing the base of the plant, grey-green, 10-50 cm long and 5-12 mm wide. **Flowers** Green and red, tubular, 6-10 cm long with 6 curled-back lobes on one side of the flower, covered in dense green hair, except at the base where it is red. They are arranged in condensed terminal racemes on red-hairy stems 30-100 cm tall. **Fruits** 3-celled capsules. **Flowering** Mainly in winter and spring. **Habitat** Sandy soils in plains, low woodlands and forests on the west coast of WA and southeastern Qld **Family** Haemodoraceae.

Macropidia fuliginosa
Black Kangaroo Paw

Upright perennial herb to 1.3 m high. **Leaves** Strap-like, arising from and sheathing the base of the plant, bluish-green, 20-50 cm long and 10-15 mm wide. **Flowers** Black and green, tubular, 5-6 cm long with 6 curled-back lobes on one side of the flower, covered in black hairs outside, clustered on branched stems. **Fruits** 3-celled capsules containing seeds 3-5 mm diameter. **Flowering** Spring. **Habitat** Gravelly heaths and open mallee woodlands of southwestern WA. **Family** Haemodoraceae.

Jasminum suavissimum (syn. J. simplicifolium)   Sweet or Spicy Jasmin

Scrambling shrub to 0.5 m high with trailing or twining branches to 3 m long. **Leaves** Opposite, rarely alternate, narrow-elliptic to linear, pointed, 15-60 mm long and 2-9 mm wide. **Flowers** White, perfumed, tubular, 8-15 mm long usually with 7-8 spreading, pointed lobes, 7-9 mm long, arranged in terminal clusters of 1-7 flowers on long stalks. **Fruits** Black shiny berries, 6-7 mm diameter. **Flowering** Spring and summer. **Habitat** Widespread in sclerophyll forests and woodlands of the coast and ranges in southeastern Qld and northern NSW. **Family** Oleaceae.

Bougainvillea spectabilis
Woolly Leaf Bougainvillea

Climbing shrub with slender branches. **Leaves** Alternate, ovate to elliptic with downy hairs, 4-10 cm long and 2-6 cm wide. **Flowers** Red-purple to scarlet, about 5 cm across, comprising 3 ovate petal-like bracts 3-6 cm long, surrounding one or 3 tubular flowers 16-24 mm long. They are arranged in leafy terminal clusters. **Flowering** Spring, summer and autumn. **Habitat** Introduced from South America, common in Qld, NSW and Vic. **Family** Nyctaginaceae.

Jasminum suavissimum

Bougainvillea spectabilis

Macropidia fuliginosa

Anigozanthos flavidus

Epacris obtusifolia

Anigozanthos manglesii

Epacris longiflora
Fuchsia Heath

Straggling shrub to 2.5 m high with wiry stems and hairy branches. **Leaves** Stalkless, ovate to heart-shaped or lanceolate, minutely-toothed, rigid with sharp points, 5-17 mm long and 3-6 mm wide. **Flowers** Red with brown or white tips, tubular, 12-40 mm long with 5 small, curled-back lobes. They are solitary, arising from the leaf axils, extending down one side of the leafy branches. **Fruits** Capsules 3-4 mm long. **Flowering** Most of the year. **Habitat** Sandy sites in heaths and woodlands along the coast and tablelands from southeastern Qld to central eastern NSW. **Family** Epacridaceae.

Coprosma nitida
Shining Coprosma

Erect or prostrate shrub to 3 m high with stiff, often spiny branches. **Leaves** Opposite, narrow-ovate to narrow-lanceolate or linear, shiny-green above, paler below, thick, with curved-back margins, 5-20 mm long and 2-5 mm wide. **Flowers** White to greenish, tubular, 2-4 mm long with 4-5 curled-back lobes and 4 long, protruding, yellow-tipped stamens in the male flowers, and 2 long protruding stigmas in the females. They are terminal and solitary on short axillary stalks. **Fruits** Oval, shining, orange-red berries about 1 cm long. **Flowering** Spring and summer. **Habitat** Subalpine forests above 900 m in the northern tablelands of NSW, Vic. and Tas. **Family** Rubiaceae.

Lambertia inermis
Chittick

Erect, bushy shrub to 6 m high with widely-spreading branches and silky new growth. **Leaves** Opposite, ovate to obovate or linear, leathery, silky-hairy below, 6-24 mm long and 3-7 mm wide. **Flowers** Yellow to orange or red, tubular, 45-55 mm long with 4 curled-back lobes and long protruding styles, arranged in stalkless terminal clusters of 6-7 flowers. **Fruits** Beaked, ovoid, woody follicles, 8-10 mm long and 6-8 mm wide with 2 winged seeds about 8 mm long. **Flowering** Year round. **Habitat** Sandy heaths and mallee woodlands in southwestern WA. **Family** Proteaceae.

Persoonia confertiflora
Cluster-flower Geebung

Erect, spreading shrub to 2 m high with angular, reddish, hairy, young branches. **Leaves** Alternate to opposite, ovate to narrow-elliptic, leathery, dark-green above, paler below with a prominent midrib and curled-back margins, 3-10 cm long and 1-3 cm wide, hairy when young. **Flowers** Yellow with rusty-brown hairs, tubular, 15-20 mm long with 4 curled-back lobes, arranged in axillary clusters of 6-10 flowers. **Fruits** Small, fleshy, greenish ovoid drupes. **Flowering** Spring to autumn. **Habitat** Wet sclerophyll forests and woodlands of the coast and ranges in southeastern NSW and eastern Vic. **Family** Proteaceae.

Persoonia juniperina
Prickly Geebung

Erect to spreading, bushy shrub to 2 m high with hairy young branches. **Leaves** Alternate, flat and stiff, narrow-linear, 8-30 mm long and 1-3 mm wide, with sharp points, densely-hairy when young, slightly channelled above. **Flowers** Yellow, tubular, 9-13 mm long with 4 curled-back lobes and scattered white hairs outside. They are solitary in the leaf axils. **Fruits** Green to indigo, succulent, ovoid drupes, 6-10 mm long. **Flowering** Summer. **Habitat** Sandy soils in heaths and open forests of southeastern NSW, Vic., southeastern SA, Vic. and Tas. **Family** Proteaceae.

Persoonia pinifolia
Pine-leaf Geebung

Tall, bushy, often drooping shrub to 4 m high with hairy young branches. **Leaves** Crowded, needle-like, 3-7 cm long and less than 1 mm wide, grooved below, hairy when young. **Flowers** Yellow, tubular, silky-hairy, to 1 cm long with 4 curled-back lobes, arranged in a dense terminal spike, 5-15 cm long. **Fruits** Green to red, succulent, ovoid drupes, 7-8 mm long. **Flowering** Spring and summer. **Habitat** Sandy soils in dry sclerophyll forests and heaths of central eastern NSW. **Family** Proteaceae.

Lambertia inermis

Persoonia pinifolia

Epacris longiflora

Persoonia juniperina

Persoonia confertiflora

Coprosma nitida

Lambertia formosa
Mountain Devil. Honey Flower

Erect, bushy shrub to 2 m high. **Leaves** Usually in whorls of 3, rigid, linear, wedge-shaped or linear-lanceolate, sharply-pointed, 2-8 cm long and 2-5 mm wide with curved-back margins, shiny-green above and whitish below, with a prominent midrib. **Flowers** Red, tubular, bearded inside, 3-5 cm long with curled-back lobes and long, straight, protruding styles. They are arranged in terminal clusters of usually 7 flowers, enclosed in red bracts. **Fruits** 2-valved, stalkless, woody follicles, 15-25 mm long, with a short beak and long horn on each valve. **Flowering** Most of the year. **Habitat** Sandy soils in heaths and dry forests of the coast and ranges of eastern NSW. **Family** Proteaceae.

Styphelia laeta
Fivecorners

Upright or spreading shrub to 2 m high with rigid, velvety branches. **Leaves** Crowded, broad-ovate to lanceolate, sharply-pointed, stiff, concave, stalkless or with very short stalks, 12-35 mm long and 3-15 mm wide. **Flowers** Pale yellow-green or red, tubular with tufts of hair inside, 14-27 mm long with 5 curled-back lobes and long, protruding stamens, usually solitary in the leaf axils. **Fruits** 5-cornered drupes, 6-9 mm long. **Flowering** From late summer to early spring. **Habitat** Widespread in heaths and dry sclerophyll forests in sandy soils of the coast and tablelands of central eastern NSW. **Family** Epacridaceae.

Styphelia longifolia
Long-leaved Fivecorners

Erect shrub to 2 m high with silky branchlets. **Leaves** Stiff, lanceolate, concave, sharply-pointed, stalkless or with short stalks, 2-5 cm long and 2-6 mm wide. **Flowers** Greenish-yellow, tubular, 18-25 mm long with 5 curled-back lobes, hairy inside, with long, protruding stamens, usually solitary in the leaf axils. **Fruits** 5-cornered drupes, 6-8 mm long. **Flowering** Winter and spring. **Habitat** Sandy soils in dry sclerophyll forests of the coast and tablelands of central eastern NSW. **Family** Epacridaceae.

Styphelia triflora
Pink Fivecorners

Erect shrub to 2 m high. **Leaves** Stiff, narrow or broad-lanceolate to obovate, 13-35 mm long and 3-9 mm wide, sharply-pointed, stalkless or with short stalks. **Flowers** Pink to red, sometimes pale-green or yellow-green, tubular, 13-30 mm long with 5 curled-back lobes, hairy inside, with long, protruding stamens, solitary or rarely in small clusters in the leaf axils. **Fruits** Finely-ribbed, 5-cornered drupes, 6-8 mm long. **Flowering** Mainly in winter and early spring. **Habitat** Widespread in sandy heaths and sclerophyll forests and woodlands of the coast and tablelands from southeastern Qld to southeastern NSW. **Family** Epacridaceae.

Styphelia tubiflora
Red Fivecorners

Upright or spreading shrub to 1 m high with downy branchlets. **Leaves** Stiff, oblong to linear, sharply-pointed, flat or convex above, 7-24 mm long and 1-4 mm wide, with very short stalks. **Flowers** Red, sometimes pale-cream or rarely white, tubular, 14-25 mm long with 5 curled-back lobes, hairy inside, with long, protruding stamens, usually solitary in the leaf axils. **Fruits** Rounded 5-cornered drupes about 5 mm long. **Flowering** Winter. **Habitat** Widespread in sandy heaths and dry sclerophyll forests of the coast and tablelands of central eastern and southeastern NSW. **Family** Epacridaceae.

Isopogon ceratophyllus
Horny Conebush

Low, spreading, dense, prickly shrub to 60 cm or rarely 1.2 m high, with red-brown branchlets. **Leaves** Rigid, 4-10 cm long with stalks 12-56 mm long, much divided into flat, narrow-linear, sharply-pointed segments 1-13 mm long. **Flowers** Yellow, tubular, 1-2 cm long with 4 curled-back lobes and long, protruding styles, arranged in dense, cone-like, terminal or axillary spikes, 15-30 mm across, surrounded by leaves. **Fruits** Dry, globular cones to 22 mm diameter, with small, hairy, beaked, ovoid nuts, 2-3 mm long. **Flowering** Winter, spring and early summer. **Habitat** Sandy soils in heaths, sclerophyll forests and woodlands of southwestern Vic., southeastern SA and northeastern Tas. **Family** Proteaceae.

Styphelia tubiflora

Styphelia triflora

Styphelia laeta

Styphelia longifolia

Isopogon ceratophyllus

Lambertia formosa

M.Westmacott

Angophora hispida
Dwarf Apple. Scrub Apple

Erect, spreading shrub or small tree to 7 m high, with dense red hairs on the young branches and flower stems. **Leaves** Opposite, leathery, stem-clasping, heart-shaped to ovate or oblong, 5-10 cm long and 30-45 mm wide, stalkless, whitish below. **Flowers** Cream, about 2 cm across with 5 small, spreading lobes and a central disc surrounded by numerous long stamens in several whorls. They are arranged in compact, globular, terminal heads on stalks with stiff red hairs. **Fruits** Cup-shaped ribbed capsules, 16-25 mm long and 14-20 mm across. **Flowering** Spring and summer. **Habitat** Locally common on sandstone in the Sydney region of NSW. **Family** Myrtaceae.

Kunzea ambigua
Tick Bush. White Kunzea

Erect shrub to 3.5 m high with hairy lateral young branchlets. **Leaves** Crowded, alternate, linear to narrow-lanceolate, concave, 4-12 mm long and 1-2 mm wide, dark-green, dotted with small glands. **Flowers** White, honey-scented, nearly stalkless, 3-4 mm across with 5 small lobes and numerous protruding stamens, 3-5 mm long. They are arranged in dense, globular or cylindrical heads on leafy branchlets, or in the axils of the upper leaves. **Fruits** Globular capsules with spiky valves, 3-4 mm diameter. **Flowering** Spring and summer. **Habitat** Widespread in sandy heaths and open forests of the coast and tablelands of, NSW, southeastern Vic. and northeastern Tas. **Family** Myrtaceae.

Kunzea capitata

Erect, twiggy shrub to 2 m high with hairy young branchlets. **Leaves** Alternate, rigid, concave, stalkless, ovate to narrow-elliptical, 3-9 mm long and 1-5 mm wide, distinctly one or 3-veined. **Flowers** Purple-pink, rarely white, 2-3 mm across with 5 small lobes and numerous protruding stamens, 3-5 mm long. They are arranged in small, compact, globular, leafless, terminal heads. **Fruits** Globular capsules with spiky valves, about 3 mm diameter. **Flowering** Late winter and spring. **Habitat** Widespread in damp sites in sandy heaths and open forests of southeastern Qld, central eastern and southeastern NSW. **Family** Myrtaceae.

Kunzea parvifolia
Violet Kunzea

Erect shrub to 1.5 m high with slender, wiry branches and downy young stems. **Leaves** Very small, alternate, almost stalkless, concave with curled-back tips, oblong to linear or lanceolate, 1-4 mm long and about 1 mm wide, dull-green, dotted with small glands. **Flowers** Pink-violet, rarely white, 3-6 mm across with 5 small lobes and numerous protruding stamens, 2-4 mm long. They are arranged in small, globular, terminal heads 10-15 mm across. **Fruits** Globular capsules with spiky valves, about 2 mm diameter. **Flowering** Spring and summer. **Habitat** Rocky sites at moderate elevations in heaths and open forests of southeastern Qld, NSW and Vic. **Family** Myrtaceae.

Darwinia fascicularis
Scent Myrtle

Erect or spreading shrub to 2 m high with slender branches. **Leaves** Scattered, often crowded in clusters at the ends of branchlets, linear, rigid, almost cylindrical, 8-16 mm long and up to 2 mm wide. **Flowers** White, turning red, tubular, 5-7 mm long with protruding styles, 12-18 mm long. They are in pincushion-like globular, terminal heads of 4-40 flowers. **Fruits** Nuts. **Flowering** From winter to summer. **Habitat** Shallow soils on sandy heaths and dry sclerophyll forests of the coast and adjacent ranges of southeastern Qld and central eastern NSW. **Family** Myrtaceae.

Melaleuca squamea
Swamp Honey Myrtle

Erect shrub or small tree to 3 m high with corky bark. **Leaves** Alternate, crowded, lanceolate with upcurved tips, 4-12 mm long and 1-3 mm wide. **Flowers** Pink to purple with 5 small lobes, 2-3 mm long, and numerous protruding stamens. They are arranged in globular, terminal heads 1-3 cm across. **Fruits** Stalkless, ovoid, woody capsules, 5-7 mm diameter, in clusters of 2-6. **Flowering** Spring. **Habitat** Damp heaths along the coast and adjacent ranges of northern and central NSW, Vic., southeastern SA and Tas. **Family** Myrtaceae.

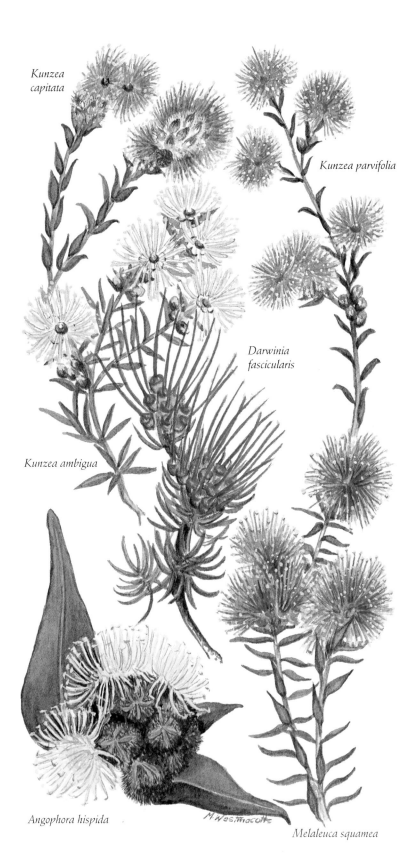

Kunzea capitata

Kunzea parvifolia

Darwinia fascicularis

Kunzea ambigua

Angophora hispida

M. Westmacott

Melaleuca squamea

Eremaea beaufortioides
Round-leaved Eremaea

Erect shrub to 3 m high with rough greyish bark and hairy young shoots. **Leaves** Alternate, stem-clasping, crowded, ovate to lanceolate with curved-back margins, 4-8 mm long and to 5 mm wide. **Flowers** Bright-orange, brush-like, about 2 cm long and 1 cm across with 5 small lobes and numerous very long, protruding stamens. They are solitary or arranged in compact, terminal, globular heads of 2-5 flowers. **Fruits** Cylindrical woody capsules about 1 cm diameter. **Flowering** Spring. **Habitat** Sandy heaths of southwestern WA. **Family** Myrtaceae.

Isopogon anemonifolius
Drumsticks

Erect or prostrate shrub to 1.5 m high. **Leaves** Stiff, flat, divided into 3 or more pointed, linear to wedge-shaped, 2 or 3 lobed segments, tipped with red, 4-11 cm long and 3-5 mm wide. **Flowers** Yellow, stalkless slender tubes, 10-12 mm long with 4 small, spreading lobes and a slightly hairy tip, arranged in dense, terminal, globular heads, 25-40 mm across with a hard central cone. **Fruits** Globular cones 10-16 mm diameter, with hairy nuts, 2-3 mm long. **Flowering** Spring and early summer. **Habitat** Sandy soils in heaths and dry forests of the coast and ranges of eastern NSW and southeastern Qld. **Family** Proteaceae.

Isopogon anethifolius
Narrow-leaf Conebush

Erect shrub to 3 m high. **Leaves** Rigid, needle-like, 4-16 cm long, divided into several pointed, needle-like segments, usually 20-35 mm long. Young growth is often reddish. **Flowers** Yellow, stalkless, slender tubes, 10-15 mm long, with 4 small, spreading lobes and short hairs, arranged in dense, terminal, ovoid to globular heads, with a hard central cone, often in clusters of 2 or 3. **Fruits** Ovoid to globular cones, 12-25 mm diameter, with hairy nuts, 5-6 mm long. **Flowering** Spring. **Habitat** Widespread in dry sclerophyll forests and heaths on sandy soils along the coast and ranges of NSW. **Family** Proteaceae.

Isopogon formosus
Rose Cone Flower

Erect, bushy, prickly shrub to 2 m high. **Leaves** Crowded, rigid, much divided into needle-like or narrow, grooved segments, with sharp points, 2-3 cm long. **Flowers** Rose-pink to red, stalkless, slender tubes to 25 mm long, with 4 small spreading lobes and a hairy tip, arranged in dense, terminal, globular heads, 4-6 cm across. **Fruits** Globular to ovoid cones, 2-3 cm diameter, with ovoid, beaked nuts to 3 mm long. **Flowering** Winter and spring. **Habitat** Sandy or gravelly depressions in southwestern WA. **Family** Proteaceae.

Isopogon latifolius

Erect shrub to 3 m high with reddish, silky, growing tips. **Leaves** Alternate, thick, variable, obovate to narrow-elliptic or lanceolate with a sharp point, light-green with prominent veins below, 4-14 cm long and 2-5 cm wide. **Flowers** Pink, stalkless, slender tubes to 4 cm long with 4 small spreading lobes and a hairy tip, arranged in dense, terminal, globular heads up to 8 cm across. **Fruits** Globular to ovoid cones, 4-5 cm diameter, with ovoid, beaked nuts to 3 mm long. **Flowering** Winter and spring. **Habitat** Sheltered rocky slopes in shrubs and mallee woodlands in the Stirling Range area of the south coast of WA. **Family** Proteaceae.

Telopea speciosissima
Waratah

Erect shrub to 3.5 m high. **Leaves** Alternate, narrow-obovate to narrow-spathulate, leathery, usually with toothed margins, 8-28 cm long and 20-65 mm wide. **Flowers** Crimson, tubular with 4 curled-back lobes and a protruding, white-tipped style, arranged in dense, compact, globular, terminal heads of 90-250 flowers, 8-15 cm diameter, surrounded by red bracts. **Fruits** Woody follicles 7-15 cm long, containing winged seeds.
Flowering Spring. **Habitat** Sandy soils with brown or yellow clay, in dry sclerophyll forests of the central and southern tablelands and south coast of NSW. **Family** Proteaceae.

Isopogon anethifolius

Eremaea beaufortioides

Isopogon anemonifolius

Telopea speciosissima

Isopogon formosus

Isopogon latifolius

Dryandra formosa
Showy Dryandra

Erect, bushy shrub or small tree, 4-8 m high, with many hairy branches. **Leaves** Alternate, narrow-linear, deeply divided into many triangular lobes with curved-back margins, soft, paler and slightly hairy below, 5-20 cm long and up to 1 cm wide. **Flowers** Shiny yellow-orange with 4 long, hairy lobes and a long, protruding style, arranged in dense, terminal, globular heads, 5-10 cm across, set in a rosette of floral leaves. **Fruits** Small capsules. **Flowering** Spring. **Habitat** Stony or peaty soils in southwestern WA. **Family** Proteaceae.

Dryandra nivea
Couch Honeypot

Low, spreading shrub with branches to 3 m long. **Leaves** Alternate, linear, deeply divided into sharply-pointed triangular lobes with curved-back margins, covered with pale hairs below, 15-40 cm long and 5-8 mm wide. **Flowers** Golden-yellow with 4 lobes and long, protruding styles, arranged in dense, terminal, globular heads, 4-5 cm across, closely surrounded by many short, hairy, floral leaves. **Fruits** Small, nut-like capsules. **Flowering** Winter and spring. **Habitat** Widespread in sandy and gravelly soils in southwestern WA. **Family** Proteaceae.

Dryandra polycephala
Many-headed Dryandra

Erect, slender shrub to 3 m high. **Leaves** Alternate, linear, rigid, divided half way to the midrib, with sharply-pointed, toothed, curved-back margins, 5-20 cm long and about 6 mm wide. **Flowers** Bright yellow with 4 lobes and long, protruding styles, arranged in many dense, terminal, globular heads, 3-4 cm across, on short branches. **Fruits** Woody follicles. **Flowering** Winter and spring. **Habitat** Sandy and gravelly soils in the Darling Range and Irwin district of southwestern WA. **Family** Proteaceae.

Dryandra quercifolia
Oak-leaf Dryandra

Erect, stiff shrub to 4 m high, with bronze hairs on the branches. **Leaves** Alternate, oblong, ovate to spathulate with irregular, sharply-pointed, toothed margins, 5-9 cm long and 4-6 cm wide. **Flowers** Yellow to greenish with 4 lobes and long, protruding styles, arranged in dense, terminal, globular heads, 6-8 cm across, surrounded by long, brown, floral leaves. **Fruits** Woody, fan-shaped capsules, about 2 cm long. **Flowering** From autumn to spring. **Habitat** Gravelly soils of the south coast of WA. **Family** Proteaceae.

Lomandra leucocephala
Woolly Mat Rush. Irongrass

Upright, perennial herb, forming dense clumps to 60 cm high. **Leaves** Long and narrow, arising from the base of the plant, 20-80 cm long and 1-5 mm wide. **Flowers** White or cream, 1-3 mm long, sweetly-perfumed, with 6 lobes and 6 yellow stamens in male flowers, arranged in dense, globular to cylindrical clusters to 5 cm across. **Fruits** Shiny capsules about 7 mm long. **Flowering** From autumn to spring. **Habitat** Widespread in sclerophyll forests on sandy soils in all mainland states. **Family** Xanthorrhoeaceae.

Acaena novae-zelandiae (syn. A. anserinifolia)
Bidgee Widgee

Sprawling, perennial herb to 15 cm high, with hairy stems and burrs that adhere to clothing and animal fur. **Leaves** Pinnately divided into 7-9 oblong to orbicular or lanceolate leaflets, 5-12 mm long and 3-5 mm wide, with toothed margins and hairs on the lower surface. **Flowers** Greenish-brown, dense, hairy, terminal, globular heads, 6-20 cm across on stalks 5-10 cm long. **Fruits** Small, dry and globular, 20-25 mm diameter with barbed spines about 1 cm long. **Flowering** Summer. **Habitat** Widespread in forests and grasslands of the coast and adjacent ranges in southeastern Qld, NSW, Vic., SA, southwestern WA and northeastern Tas. **Family** Rosaceae.

Lomandra leucocephala

Dryandra formosa

Dryandra nivea

Acaena novae-zelandiae

Dryandra quercifolia

Dryandra polycephala

Hoya australis
Australian Waxplant. Native Hoya

Slender, fleshy, climbing herb, often scrambling over rocks, with twining stems to 6 m long. **Leaves** Opposite, broad-elliptic to egg-shaped, 3-12 cm long and 2-6 cm wide, dark-green and fleshy, hairy below. **Flowers** White or pale pink, open, 10-25 mm across with 5 pointed lobes and 5 stamens with a fleshy appendage, waxy and sweetly-perfumed. They are arranged in dense, axillary, globular clusters of 10-30 flowers, about 7 cm across. **Fruits** Cylindrical to oblong follicles, 10-15 cm long and 10-15 mm wide. **Flowering** Autumn and winter. **Habitat** Dry rainforests and coastal rocky areas exposed to salt spray, from northeastern Qld to northeastern NSW. **Family** Asclepiadaceae.

Pimelea ferruginea
Pink Rice Flower

Dense, erect shrub to 1.5 m high. **Leaves** Opposite, crowded, oblong to ovate or elliptic with curved-back margins, shiny, stalkless, 5-16 mm long and 1-6 mm wide. **Flowers** Pink, sometimes white, tubular, 7-15 mm long, with 4 spreading lobes, silky-hairy with 2 protruding stamens, arranged in dense, terminal, globular heads, 3-4 cm across. **Fruits** Drupes with oval nuts about 3 mm long. **Flowering** Winter, spring and summer. **Habitat** Sand dunes and rocky outcrops along the southwestern coast of WA. **Family** Thymelaeaceae.

Pimelea ligustrina
Tall Rice Flower

Erect or spreading shrub to 3 m high with slender branches. **Leaves** Opposite in pairs, narrow-elliptic to lanceolate with slightly curved-back margins, shiny-green above, paler below, thin and soft, 1-9 cm long and 2-28 mm wide. **Flowers** Creamy-white, tubular, hairy outside, 6-17 mm long with 2 protruding stamens, arranged in terminal, dense, globular heads of 50-130 flowers, 2-3 cm across, surrounded by floral leaves. **Fruits** Green drupes, 3-5 mm long, with ovoid nuts. **Flowering** Spring and summer. **Habitat** Widespread in moist forests of the coast and ranges in southeastern Qld, NSW, Vic., southeastern SA and Tas. **Family** Thymelaeaceae.

Pimelea linifolia
Slender Rice Flower

Erect or prostrate shrub to 1.5 m high, toxic to stock. **Leaves** Opposite, lanceolate to narrow-elliptic with a prominent midrib, 3-32 mm long and 1-7 mm wide. **Flowers** Creamy-white tinged with pink, tubular, hairy outside, 1-2 cm long with 4 spreading lobes and 2 protruding stamens, arranged in globular, terminal heads of 7-60 flowers, 3-6 cm across, surrounded by floral leaves. **Fruits** Green, ovoid drupes, 3-5 mm long. **Flowering** Most of the year. **Habitat** Widespread in heaths and forests in Qld, NSW, Vic., SA and Tas. **Family** Thymelaeaceae.

Pimelea suaveolens
Silky-yellow Banjine

Erect, slender shrub to 1.2 m high, often multi-stemmed at the base. **Leaves** Opposite, narrow-lanceolate to narrow-linear, dull-green, slightly hairy, concave, usually stalkless, 5-34 mm long and 2-6 mm wide. **Flowers** Yellow, tubular, slightly hairy outside, 8-15 mm long with 4 spreading lobes and 2 protruding stamens, arranged in terminal, dense, globular heads, 3-4 cm across, surrounded by a rosette of 4-8 large green floral leaves. **Fruits** Drupes. **Flowering** Winter and spring. **Habitat** Widespread in clay or sandy soil along the coast and inland from the Darling Range to Albany in southwestern WA. **Family** Thymelaeaceae.

Pimelea sylvestris

Erect, wiry shrub to 2 m high. **Leaves** Opposite in pairs, variable, oblong to elliptic or lanceolate, sometimes lobed, 12-45 mm long and 2-20 mm wide. **Flowers** Pink and white, tubular, 5-8 mm long with 4 spreading lobes and 2 protruding stamens, arranged in terminal globular heads about 2 cm across, surrounded by a rosette of 4-6 green floral leaves. **Fruits** Drupes with nut-like seeds. **Flowering** Spring. **Habitat** Widespread on sandy heaths, gullies and hills in woodlands and forests of the southwestern coast and interior of WA. **Family** Thymelaeaceae.

Pimelea ligustrina

Pimelea linifolia

Hoya australis

Pimelea sylvestris

Pimelea ferruginea

Pimelea suaveolens

Acacia myrtifolia
Myrtle or Red-stemmed Wattle

Erect or prostrate shrub to 3 m high with smooth grey bark and angled or flattened branchlets. **Leaves** (phyllodes) Alternate, narrow-elliptic with a short point, 2-6 cm long and 5-30 mm wide, leathery with a prominent central vein and thickened yellow margins. Flowerheads are pale-yellow to white fluffy balls of 2-8 flowers, 6-10 mm diameter, on short axillary stalks. **Fruits** Brown, woody, narrow, curved pods with thick margins, 4-11 cm long and 3-5 mm wide, slightly constricted between the seeds. **Flowering** Winter and spring. **Habitat** Widespread on sandy soils in heaths and open forests along the coast and ranges of southeastern Qld, NSW, Vic., southeastern SA, southwestern WA and Tas. **Family** Mimosaceae.

Acacia paradoxa *(syn. A. armata)*
Kangaroo Thorn. Hedge Wattle

Straggling shrub to 4 m high with finely-fissured, brownish-grey bark and sharp thorns along the angular or flattened branchlets. **Leaves** (phyllodes) Alternate, pointed, straight or sickle-shaped, usually 8-30 mm long and 3-7 mm wide, pointed, dark-green and leathery with wavy margins and a prominent central vein. **Flowerheads** Fluffy, lemon-yellow clusters of 20-45 flowers, about 8 mm diameter, solitary on short axillary stalks. **Fruits** Straight or curved pods, 2-7 cm long and 3-5 mm wide with thickened margins, brown, often covered in whitish furry hairs. **Flowering** Winter and spring. **Habitat** Widespread in many different communities in all states except the NT. **Family** Mimosaceae.

Acacia rigens
Nealie. Needle Wattle

Rounded, erect or spreading shrub to 4 m high with dark-grey, fissured bark and angular or flattened branchlets. **Leaves** (phyllodes) Alternate, needle-like with a sharp point, often curved, 3-15 cm long and 1-3 mm wide, greyish-green, tough and flexible. **Flowerheads** Fluffy, golden-yellow balls of 20-30 flowers, about 5 mm diameter, on short axillary stalks, solitary or in clusters of 2-4 flowers. **Fruits** Brown, hairy, curved, twisted or coiled pods, 4-12 cm long and 2-8 mm wide, constricted between the seeds. **Flowering** Winter and spring. **Habitat** Drier sandy soils in mallee scrubs and woodlands, inland in southern Qld, NSW, western Vic., SA and southern WA. **Family** Mimosaceae.

Acacia stricta
Hop Wattle

Erect or spreading shrub or small tree to 6 m high with smooth bark and angular or flattened branchlets. **Leaves** (phyllodes) Lanceolate to narrow-linear, rarely with a short point, 5-15 cm long and 3-15 mm wide with a prominent central vein. **Flowerheads** Fluffy, pale-yellow to white balls of 20-30 flowers, about 8 mm diameter on short stalks, arranged in axillary clusters of 1-4 flowers. **Fruits** Pale-brown, rough, usually flat, narrow pods, 3-10 cm long and 2-5 mm wide. **Flowering** Winter and spring. **Habitat** Widespread on moist sites in open forests of the coast and ranges in southeastern Qld, NSW, Vic., southeastern SA and northeastern Tas. **Family** Mimosaceae.

Acacia suaveolens
Sweet Wattle

Stiff, slender, erect to prostrate shrub to 2.5 m high with smooth, purplish-brown bark and angular or flattened branchlets. **Leaves** (phyllodes) Alternate, narrow-oblong to lanceolate, leathery, 5-15 cm long and 2-10 mm wide with a prominent central vein and thickened margins. **Flowerheads** Fluffy, pale-yellow to white balls of 3-10 flowers, scented, about 6 mm diameter on short stalks, arranged in short axillary racemes. **Fruits** Pale bluish-green, usually flat, oblong pods, 2-5 cm long and 8-20 mm wide. **Flowering** Autumn, winter and spring. **Habitat** Sandy soils in heaths and open forests along the coast and ranges of southeastern Qld, NSW, Vic., southeastern SA and Tas. **Family** Mimosaceae.

Acacia ulicifolia
Prickly Moses. Juniper Wattle

Erect or spreading prickly shrub to 2 m high with smooth grey bark and drooping branches. **Leaves** (phyllodes) Alternate, crowded, needle-like, stiff and sharp, dark-green, 8-15 mm long and 1-2 mm wide with a prominent central vein. **Flowerheads** Fluffy, creamy-yellow balls of 15-35 flowers, about 7 mm diameter, usually solitary on slender axillary stalks. **Fruits** Flat, narrow, curved pods, 2-6 cm long and 3-5 mm wide with thickened margins, constricted between the seeds. **Flowering** Winter and spring. **Habitat** Poorly drained, sandy sites, in heaths and open forests of the coast and tablelands in Qld, NSW, Vic. and northeastern Tas. **Family** Mimosaceae.

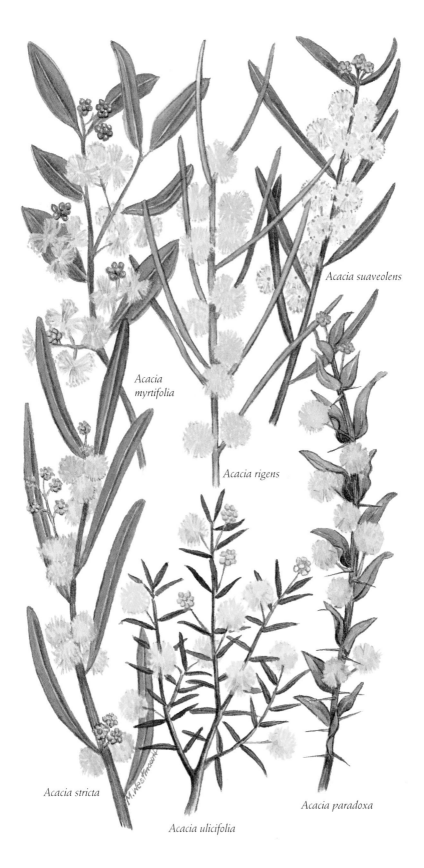

Acacia suaveolens

Acacia myrtifolia

Acacia rigens

Acacia stricta

Acacia paradoxa

Acacia ulicifolia

Brunonia australis
Blue or Australian Pincushion
Upright perennial herb to 1 m high, covered with soft hair. **Leaves** Elliptic to spathulate, pointed, silky-hairy, arising from the base of the stem, 4-15 cm long and 2-18 mm wide. **Flowers** Blue, tubular, 3-4 mm long with 5 spreading lobes and 5 protruding, yellow-tipped stamens. They are arranged in solitary, terminal, dense, globular heads, 13-25 mm diameter, on hairy stalks, 5-30 cm long. **Fruits** Nuts about 3 mm long. **Flowering** Mainly in spring. **Habitat** Widespread in dry sclerophyll forests, woodlands and sand dunes, inland in all states. **Family** Goodeniaceae/Brunoniaceae.

Acacia continua
Thorn Wattle
Erect or spreading prickly shrub to 2 m high with angular or flattened, densely-hairy, often intertwined branchlets. **Leaves** (phyllodes) Alternate, stalkless, needle-like with a sharp point, rigid, 1-4 cm long and about 1 mm wide. **Flowerheads** Fluffy, bright-yellow balls of about 30 flowers, about 1 cm across, solitary or in pairs on short axillary stalks. **Fruits** Curved or twisted pods, 3-8 cm long and 3-5 mm wide, constricted between the seeds. **Flowering** Winter and spring. **Habitat** Dry rocky ridges, in mallee and cypress pine woodlands, inland in NSW and SA. **Family** Mimosaceae.

Acacia ligulata
Dune Wattle. Umbrella Bush
Erect or spreading bushy shrub to 4 m high, with angled or flattened branchlets. **Leaves** (phyllodes) Alternate, thick, linear to oblong, 1-veined, grey-green, 3-10 cm long and 2-10 mm wide. **Flowerheads** Bright-yellow to orange fluffy balls of 15-20 flowers, about 6 mm across, solitary or 2-5 together in short racemes on a short flower stalk. **Fruits** Hard, brown, straight to curved pods, 3-10 cm long and 5-10 mm wide, constricted between the seeds. **Flowering** Usually in winter and spring. **Habitat** Widespread in mulga and bluebush communities and woodlands, often on sand dunes, inland in all mainland states. **Family** Mimosaceae.

Acacia terminalis
Sunshine Wattle
Erect or spreading shrub, rarely a small tree, 1-5 m high with angular, deep-red young branches. **Leaves** Alternate, 3-8 cm long, bipinnately divided into 2-6 pairs of segments, 3-5 cm long, each with 8-16 pairs of blunt lanceolate leaflets, paler below, 8-20 mm long and 2-5 mm wide. **Flowerheads** Pale to golden-yellow or white fluffy balls of 6-15 flowers, about 6 mm across, in terminal spikes. **Fruits** Wrinkled, oblong, slightly curved or straight pods, reddish when young, flat, 3-11 cm long and 8-17 mm wide with thickened margins. **Flowering** From late summer to spring. **Habitat** Widespread in scrub and dry open forests along the coast and tablelands of NSW, Vic., western and southwestern Tas. **Family** Mimosaceae.

Trachymene coerulea
Rottnest Daisy. Blue Lace-flower
Upright annual or biennial herb to 60 cm high with coarse hairy stems. **Leaves** Deeply dissected into 3-5 segments each with 3 pointed lobes, up to 4 cm long, arising from the base of the plant. **Flowers** Blue to white with 5 lobes and 5 stamens, arranged in dense globular heads, 3-5 cm across, surrounded by floral leaves. **Fruits** Compressed discs 4-6 mm across. **Flowering** Summer. **Habitat** Limestone sites in southwestern WA. **Family** Apiaceae.

Lantana camara
Lantana
Scrambling shrub with prickly branches to 5 m long, forming dense thickets. **Leaves** Opposite, oblong to ovate or slightly heart-shaped, rough with toothed margins, hairy below, 2-12 cm long and 15-45 mm wide. **Flowers** Pale-yellow, orange or pink turning deep-red, tubular, 6-14 mm long with 4 spreading lobes, arranged in dense axillary globular heads, 2-5 cm across. **Fruits** Small, fleshy, dark-purple drupes, 4-6 mm diameter. **Flowering** Most of the year. **Habitat** Introduced from tropical South America. A widespread weed on fertile soils in disturbed sites in rainforests and sclerophyll forests of coastal Qld, NSW, Vic., southern SA, northern NT and southwestern WA. **Family** Verbenaceae.

Acacia ligulata

Acacia terminalis

Trachymene coerulea

Acacia continua

Brunonia australis

Lantana camara

Calocephalus brownii
Cushion or Snow Bush

Low, hardy, bushy shrub to 2 m high with densely entangled, grey-white branches. **Leaves** Very narrow, 2-15 mm long, downy, grey, and held close to the branches. **Flowerheads** Small, whitish-yellow, compact, composite, covered in woolly hairs and packed into globular, terminal flowerheads, 8-12 mm diameter. **Flowering** Spring and summer. **Habitat** Exposed faces of cliffs and dunes along the coast of Vic., SA, south-western WA and northeastern Tas. **Family** Asteraceae.

Ozothanmus turbinatus
Coast Everlasting
(syn. Helichrysum paralium)

Dense, erect shrub to 2.5 m high with white hairy branches. **Leaves** Linear, stiff, crowded around the branches, 10-25 mm long and 1-2 mm wide with margins tightly rolled under, smooth green above with a depressed central vein and white woolly below, usually tinged yellow at the base. **Flowerheads** Whitish-yellow, small, compact, composite buttons, 3-7 mm wide. They are arranged in crowded terminal clusters of 15-30 flowers. **Fruits** Dark-brown oblong achenes with yellow hairs. **Flowering** Summer and autumn. **Habitat** Exposed cliffs and sand dunes along the coast of southern NSW, Vic., southeastern SA and Tas. **Family** Asteraceae.

Ozothanmus obcordatus
Grey Everlasting
(syn. Helichrysum obcordatum)

Erect, slender shrub to 2 m high with greyish, woolly young branches. **Leaves** Alternate, broad-obovate to broad-elliptic, stiff with curled-under margins, 3-15 mm long and 2-9 mm wide, pale below with white or grey hairs. **Flowerheads** Yellow, compact, composite buttons, 3-4 mm long and 1-2 mm across, arranged in compact, flat-topped, terminal clusters, 2-10 cm wide. **Fruits** Angular achenes. **Flowering** Spring and summer. **Habitat** Shallow stony soils in heaths of the coast, ranges and inland slopes of southeastern Qld, NSW, Vic. and northeastern Tas. **Family** Asteraceae.

Ozothanmus diosmifolius (syn. Helichrysum diosmifolium)

Erect, straight-stemmed shrub to 5 m high with minutely hairy branches. **Leaves** Narrow-linear with a small hooked tip and rolled-under margins, 8-25 mm long and 1-2 mm wide, shiny-green and rough above, white and woolly below. **Flowerheads** White or pink, small, compact, composite buttons, 2-3 mm long and 2-3 mm wide, arranged in dense, broad, flat-topped, terminal clusters. **Fruits** Achenes. **Flowering** Late winter to spring. **Habitat** Open forests, rainforest margins and heaths, often on hillsides along the east coast, ranges and inland slopes of southeastern Qld and NSW. **Family** Asteraceae.

Senecio odoratus
Scented Groundsel

Coarse, erect under-shrub to 2 m high. **Leaves** Alternate, narrow to broad-lanceolate or obovate, often with curved-back and toothed margins, thick and fleshy, 4-14 cm long and 5-55 mm wide, dark-green with a whitish coating above, sometimes cottony below. **Flowerheads** Yellow, composite, cylindrical buttons, 5-9 mm long and about 5 mm across. They are arranged in terminal clusters. **Fruits** Yellow to brown achenes about 2 mm long. **Flowering** Summer. **Habitat** Rocky locations in western Qld, coastal Vic., southeastern SA and coastal Tas. **Family** Asteraceae.

Craspedia uniflora
Common Billy's Buttons

Erect, annual or perennial, tufted herb, to 50 cm high. **Leaves** Oblong or broad-lanceolate to narrow-elliptic, arising from the base of the plant, sparsely hairy, 5-25 cm long and 1-3 cm wide. **Flowerheads** Pale-yellow to orange composite buttons, ovoid or almost globular, 15-35 mm across, solitary and terminal, with narrow leafy bracts along the stem. **Flowering** Spring and summer. **Habitat** Widespread along the coast and tablelands in southern Qld and southern SA. **Family** Asteraceae.

*Ozothamnus
obcordatus*

Ozothamnus turbinatus

Calocephalus brownii

Craspedia uniflora

Senecio odoratus

*Ozothamnus
diosmifolius*

Chrysocephalum apiculatum
Yellow Buttons. Common Everlasting
(syn. Helichrysum apiculatum)

Variable perennial herb, 7-60 cm high, with woolly stems. **Leaves** Linear to lanceolate or wedge-shaped, soft, 1-6 cm long and 10-25 mm wide, covered with soft silvery hairs. **Flowerheads** Composite buttons, bright-yellow, sometimes tinged with pink, 5-8 mm long and 7-15 mm across, arranged in dense terminal clusters. **Fruits** Oblong to ovoid, 4-angled achenes. **Flowering** Mainly in spring. **Habitat** Widespread in various communities, often in disturbed and open sites of all states. **Family** Asteraceae.

Helichrysum rutidolepis
Yellow Paper Daisy. Pale Everlasting

Prostrate perennial herb with turned-up stems, 15-40 cm high, with woolly branches. **Leaves** Linear to lanceolate with curved-back margins, often with a sharp point, 25-80 mm long and 1-8 mm wide, usually with woolly hairs below or on both surfaces. **Flowerheads** Yellow, papery, composite buttons, 10-20 mm across and 6-12 mm long, solitary and terminal on almost leafless stems. **Fruits** Brown, compressed, oblong achenes. **Flowering** Mostly in summer and autumn. **Habitat** Grasslands and wet sclerophyll forests and woodlands of the coast and adjacent ranges of NSW, Vic. and southeastern SA. **Family** Asteraceae.

Myriocephalus stuartii
Poached Egg Daisy

Erect, woolly, sticky, annual herb, to 60 cm high. **Leaves** Linear to lanceolate, grey-green and woolly with a prominent midrib, 2-7 cm long and 1-5 mm wide. **Flowerheads** White composite buttons with a large yellow centre, 2-4 cm across, papery, solitary and terminal. **Fruits** Narrow, obovoid achenes, 2-3 mm long, silky-hairy. **Flowering** Spring. **Habitat** Sand dunes and sand plains in the western plains of Qld and NSW, western Vic., SA, the central west coast of WA and southern NT. **Family** Asteraceae.

Cotula coronopifolia
Water Buttons

Erect to spreading perennial herb to 30 cm high with fleshy stems. **Leaves** Linear to oblong, fleshy and stem-clasping, sometimes coarsely-toothed, 2-6 cm long. **Flowerheads** Yellow, composite, flattish buttons, 5-12 mm across, terminal and solitary on long stalks. **Fruits** Flattened, hairy achenes, 1-2 mm long. **Flowering** Winter and spring. **Habitat** Native of South Africa, widespread in wet sites, flooded hollows and disturbed saline communities along the coast, ranges and inland slopes of southeastern Qld, NSW, Vic., SA, southwestern WA and northeastern Tas. **Family** Asteraceae.

Leptorhynchos squamatus
Scaly Buttons

Upright perennial herb to 40 cm high with shiny reddish stems. **Leaves** Alternate, oblong to lanceolate, 15-35 mm long and 2-4 mm wide, usually with woolly hairs below. **Flowerheads** Yellow composite buttons, almost hemispherical, 7-10 mm long and 8-15 mm across, sheathed in dry woolly bracts, solitary and terminal on long stems. **Fruits** 4-angled achenes withe fine hairs. **Flowering** Summer and autumn. **Habitat** Often at higher altitudes in open areas and grasslands of NSW, Vic., SA and Tas. **Family** Asteraceae.

Decazesia hecatocephala

Upright annual herb with downy stems to 30 cm high. **Leaves** Succulent, ovate to oblong or lanceolate, woolly, grey-green, to 4 cm long and 1 cm wide. **Flowerheads** Yellow, composite, scented buttons, surrounded by white papery bracts, 10-15 mm across, solitary and terminal on thick stalks. **Flowering** Late winter and spring. **Habitat** Mulga scrub country along the coast and interior of northern WA. **Family** Asteraceae.

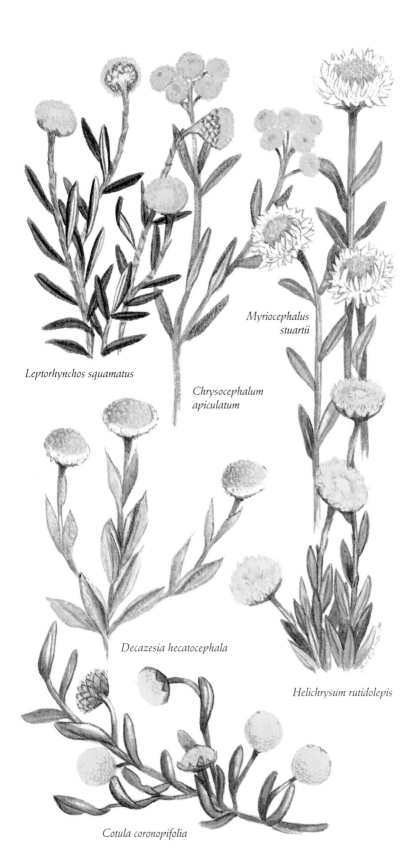

Leptorhynchos squamatus

Myriocephalus stuartii

Chrysocephalum apiculatum

Decazesia hecatocephala

Helichrysum rutidolepis

Cotula coronopifolia

Bracteantha bracteata
(syn. *Helichrysum bracteatum*)

Straw Flower. Golden Everlasting

Upright, annual or perennial herb, to 80 cm high, usually covered with small hairs. **Leaves** Stem-clasping, soft, linear to oblong or lanceolate, often hairy, 2-11 cm long and 5-40 mm wide. **Flowerheads** Golden-yellow and glossy, composite, daisy-like, papery, 25-30 mm wide and 1-2 cm long, solitary or a few together on separate stalks. **Fruits** Dark-brown, oblong, barbed achenes, about 4 mm long. **Flowering** Mainly in spring. **Habitat** Widespread, particularly in tall forests and woodlands on sandy soils in all mainland states. **Family** Asteraceae.

Helichrysum elatum

White Paper Daisy

Tall, woody shrub to 2 m high. The branches are covered in woolly hairs. **Leaves** Alternate, lanceolate, arising from the base of the stem, 8-12 cm long and 15-40 mm wide, white-woolly below. **Flowerheads** White, sometimes tinged with pink, with a yellow centre, papery, composite, daisy-like, 25-40 mm wide and 20-25 mm long with 4-8 heads in a terminal, leafy cluster. **Fruits** 4-angled achenes with white bristles. **Flowering** Late winter and spring. **Habitat** Wet and dry sclerophyll forests and rainforest margins at higher altitudes along the coast and adjacent ranges of Qld, NSW and southeastern Vic. **Family** Asteraceae.

Ozothamnus stirlingii *(syn. Helichrysum stirlingii)*

Ovens Everlasting

Open, aromatic, resinous shrub to 3 m high. **Leaves** Flat, alternate, narrow-lanceolate, 4-10 cm long and 10-15 mm wide, dark-green and smooth above, paler below with tiny hairs. **Flowerheads** White with a large brownish centre, composite, daisy-like buttons, 5-10 mm long and 8-15 mm wide, with brown, papery bracts. They are arranged in loose terminal clusters. **Fruits** Rough achenes. **Flowering** Spring and summer. **Habitat** Subalpine forests between 1000 and 1500 m in southern NSW and northeastern Vic. **Family** Asteraceae.

Waitzia acuminata

Orange Immortelle

Erect, hairy, annual herb to 60 cm high. **Leaves** Alternate, lanceolate to narrow-oblong and about 11 cm long at the base of the plant, often absent when flowering. Stem leaves are narrow-lanceolate to broad-linear with curled-under margins, 1-10 cm long and 2-5 mm wide, woolly or with soft hairs. **Flowerheads** Solitary, terminal, composite, 15-20 mm across, covered in yellow to orange stiff papery bracts. They are arranged in loose terminal clusters of 2-9 flowerheads on woolly stalks. **Fruits** Brown, beaked achenes, 3-4 mm long. **Flowering** Spring. **Habitat** Mallee and woodlands in drier sandy areas of western NSW, northwestern Vic., SA, WA and the NT. **Family** Asteraceae.

Leucochrysum albicans *(syn. Helipterum albicans)*

Hoary Sunray

Erect, woolly, perennial herb to 45 cm high. **Leaves** Alternate, linear to oblong or broad-obovate and pointed, fleshy in alpine varieties, arising from the base of the stem, often woolly-white with curved-back margins, 25-100 mm long and 1-11 mm wide. **Flowerheads** White to pale-yellow with a yellow centre, composite, daisy-like, papery, 17-40 mm across, solitary and terminal on slender stalks, 7-15 cm long. **Fruits** Brown achenes about 2 mm long. **Flowering** Summer. **Habitat** Widespread in a range of communities from the plains to wet upland areas in grasslands, heaths and forests of southern Qld, NSW, Vic. and Tas. **Family** Asteraceae.

Rhodanthe chlorocephala *(syn. Helipterum roseum)*

Pink Paper Daisy

Erect, slender, annual herb to 50 cm high. **Leaves** Alternate, linear to lanceolate, 5-15 mm long and about 2 mm wide. **Flowerheads** Pink and white with a brown centre surrounded by a yellow ring, composite, daisy-like, papery, to 5 cm across, solitary and terminal. **Fruits** Woolly achenes. **Flowering** Winter and spring. **Habitat** Arid inland areas of southwestern WA. **Family** Asteraceae.

*Waitzia
acuminata*

*Bracteantha
bracteata*

Leucochrysum albicans

Helichrysum elatum

*Ozothamnus
stirlingii*

*Rhodanthe
chlorocephala*

M. Westmacott.

Olearia asterotricha   Rough Daisy Bush

Erect shrub to 2 m high. **Leaves** Alternate, variable, usually narrow-linear to obovate, rough with rigid hairs above and softer, greyish hairs below, with curved-back and sometimes toothed or lobed margins, 6-30 mm long and 2-6 mm wide. **Flowerheads** White or pale-blue, composite, daisy-like, 12-21 mm across, solitary or clustered and terminal. **Fruits** Silky achenes. **Flowering** Spring, summer and autumn. **Habitat** Open forests and damp heaths in mountainous regions of central eastern NSW and southern Vic. **Family** Asteraceae.

Celmisia longifolia   Silver or Snow Daisy

Erect perennial herb to 30 cm high. **Leaves** Linear to lanceolate, arising from the base of the stem, with curled-under margins, 10-30 cm long and 4-16 mm wide, silver-hairy below. **Flowerheads** White or pink with a yellow centre, composite, daisy-like, 2-5 cm across, terminal and solitary on stems to 50 cm long, with leaf-like sheathing bracts 1-3 cm long. **Fruits** Hairless achenes to 6 mm long. **Flowering** Summer. **Habitat** Moist areas at higher elevations in the central and southern tablelands of NSW and Tas. **Family** Asteraceae.

Rhodanthe floribunda   White Paper Daisy. Common White Sunray
(syn. Helipterum floribundum)

Erect or prostrate, woolly, annual, rarely perennial, herb to 40 cm high. **Leaves** Alternate, linear to oblong, pointed, with curved-back margins, woolly-hairy, 5-20 mm long and 1-3 mm wide. **Flowerheads** White with a yellow centre, hemispherical, composite, daisy-like with 20-50 curled-up rays, 1-2 cm across, solitary and terminal but arranged in leafy clusters. **Fruits** Cream, top-shaped achenes about 3 mm long. **Flowering** Spring. **Habitat** Widespread on moist sites in arid inland areas of Qld, NSW, SA, WA and the NT. **Family** Asteraceae.

Argentipallium obtusifolium   Blunt Everlasting
(syn. Helichrysum obtusifolium)

Upright perennial herb to 40 cm high with stiff, greyish, hairy stems. **Leaves** Scattered, linear to narrow-oblanceolate, blunt with curled-under margins, dull above with whitish hairs below, 2-25 mm long and 1-2 mm wide. **Flowerheads** White, rarely tinged with pink, with a yellow centre, composite, daisy-like, 12-15 mm long and 2-3 cm across with short brown outer bracts, solitary and terminal on leafy branchlets. **Fruits** Angular, brown achenes. **Flowering** Spring. **Habitat** Widespread in heaths on sandy soils of southeastern SW, Vic., eastern SA, southwestern WA and Tas. **Family** Asteraceae.

Chrysanthemoides monilifera   Bitou Bush. Boneseed

Erect or sprawling shrub to 2 m high with upturned branches. **Leaves** Alternate, obovate to lanceolate with coarsely-toothed margins, often covered with a cottony substance below, leathery with a prominent midrib, 2-9 cm long and 1-5 cm wide. **Flowerheads** Yellow, composite, daisy-like, with 4-8 rays, 1-3 cm across, arranged in a terminal, flat-topped cluster of 3-12 flowers. **Fruits** Globular, purplish-black, hard or succulent drupes, 6-8 mm diameter. **Flowering** Spring. **Habitat** Introduced from South Africa, a noxious weed in Australia, growing in disturbed sites and on sand dunes in southeastern Qld, the far west plains and coastal NSW, Vic., SA, southwestern WA and Tas. **Family** Asteraceae.

Brachycome lineariloba   Dwarf Brachycome. Hard-headed Daisy

Low annual herb to 30 cm high. **Leaves** arise from the base of the plant. They are 3-8 cm long, sometimes divided into 3-9 succulent linear segments to 13 mm long. **Flowerheads** White or violet with a yellow centre, daisy-like, composite, 11-25 mm across with 8-15 rays, solitary and terminal on long stiff stalks. **Fruits** Pale brown, silky, angled, wedge-shaped achenes, 3-5 mm long. **Flowering** Winter and spring. **Habitat** Widespread in a variety of habitats, particularly woodlands and mallee communities in western NSW, western Vic., SA., and the Coolgardie area of WA. **Family** Asteraceae.

Chrysanthemoides monilifera

Argentipallium obtusifolium

*Olearia
asterotricha*

*Rhodanthe
floribunda*

Celmisia longifolia

Brachycome lineariloba

Olearia elliptica
Sticky Daisy Bush
Upright shrub to 2 m high. **Leaves** Alternate, sticky, narrow to broad-elliptic, rarely toothed, 2-12 cm long and 5-38 mm wide, paler-green below. **Flowerheads** White with a brown-yellow centre, composite, daisy-like, 6-26 mm across. They are arranged in clusters at the ends of branchlets. **Fruits** Achenes. **Flowering** From spring to autumn. **Habitat** Wet areas in forests, heaths and woodlands along the coast, tablelands and inland slopes of southeastern Qld and NSW. **Family** Asteraceae.

Olearia stellulata (syn. O. lirata)
Showy Daisy Bush
Bushy shrub to 3 m high. **Leaves** Alternate, soft, thin, linear to broad-elliptic or ovate, 5-150 mm long and 5-35 mm wide, with entire or toothed margins, smooth green above and whitish with minute hairs below. **Flowerheads** White with a yellow-brown centre, composite, daisy-like, 11-25 mm across. They are arranged in leafy, terminal panicles. **Fruits** Silky achenes. **Flowering** Spring and summer. **Habitat** Sheltered sites in sclerophyll forests and along watercourses along the coast and ranges of southeastern Qld, eastern NSW, Vic. and Tas. **Family** Asteraceae.

Olearia phlogopappa
Dusty Daisy Bush
Variable low, spreading or upright shrub to 1 m high. **Leaves** Alternate, elliptic to obovate, often with bluntly-toothed margins, 5-30 mm long and 2-10 mm wide, grey-green above and whitish with minute hairs below. **Flowerheads** Solitary or clustered, terminal, white, rarely blue, with a yellow centre, composite, daisy-like with 8-12 rays, 17-25 mm across. **Fruits** Achenes. **Flowering** Spring and summer. **Habitat** Higher elevations in sclerophyll forests, heaths and woodlands of eastern NSW, Vic. and Tas. **Family** Asteraceae.

Olearia pimeleoides
Pimelea Daisy Bush
Dense, rounded shrub to 2 m high, with greyish hairy branches. **Leaves** Alternate, obovate, linear or elliptic with curled-back margins, entire or irregularly-toothed, 3-25 mm long and 1-7 mm wide, whitish and woolly below with a prominent midrib. **Flowerheads** Solitary or clustered, terminal, white with a yellow centre, composite, daisy-like with 8-25 rays, 13-40 mm across. **Fruits** Silky, slightly flattened achenes, 2-3 mm long. **Flowering** Autumn and winter. **Habitat** Mallee scrub and drier woodlands in inland areas of southern Qld, NSW, western Vic., SA and southern WA. **Family** Asteraceae.

Olearia ramulosa
Twiggy Daisy Bush
Variable, usually a spindly, hairy shrub to 2 m high. **Leaves** Alternate, linear to narrow-elliptic or narrow-obovate with curled-back margins, 2-10 mm long and 1-3 mm wide, woolly below with a distinct midrib and sometimes sticky. **Flowerheads** White or mauve with pale-yellow centres, composite, daisy-like, 10-22 mm across, with 2-13 rays. They are solitary in long, leafy, terminal or axillary spikes. **Fruits** Pointed achenes 1-2 mm long. **Flowering** Spring, summer and autumn. **Habitat** Widespread on hilly sites in sclerophyll forests and heaths in southeastern Qld, NSW, Vic., SA and Tas. **Family** Asteraceae.

Olearia tomentosa
Toothed Daisy Bush
Upright shrub to 2 m high with stiff and furry young stems. **Leaves** Alternate, broad-ovate with bluntly-toothed or lobed margins, 18-85 cm long and 1-5 cm wide, grey or rusty-hairy below. **Flowerheads** White, blue or purple with yellow centres, borne on thick, brown, hairy stems, composite, daisy-like, 25-60 mm across with 13-29 rays, usually arranged in a terminal cluster. **Fruits** Smooth or silky achenes. **Flowering** Spring, summer and autumn. **Habitat** Widespread in dry sclerophyll forests, scrubs and heaths, on coastal headlands and rocky sites along the coast and ranges of NSW and southeastern Vic. **Family** Asteraceae.

Olearia elliptica

Olearia
stellulata

Olearia tomentosa

Olearia
pimeleoides

Olearia phlogopappa

Olearia ramulosa

Calotis cuneifolia
Bindi Eye. Purple Burr-daisy

Perennial hairy herb with ascending or prostrate stems to 60 cm high. **Leaves** Alternate, spathulate to wedge-shaped, toothed or lobed towards the tips, 8-50 mm long and 5-20 mm wide. **Flowerheads** White to lilac with a yellow centre, composite, daisy-like, 6-25 mm across with 30-45 rays. They are solitary or in small clusters on terminal or axillary stalks. **Fruits** Ridged achenes 1-2 mm long. **Flowering** Most of the year. **Habitat** Widespread in many situations including river floodplains, open forests and grasslands of the coast and tablelands of Qld, NSW, Vic. southeastern SA, central western WA and southern NT. **Family** Asteraceae.

Brachycome aculeata
Hill Daisy

Upright, tufted, perennial herb to 60 cm high. **Leaves** Spathulate to linear, toothed or lobed at the tips, 5-10 cm long and 15-20 mm wide. **Flowerheads** Pink, white, lilac or blue, daisy-like, composite, 1-2 cm across with a yellow centre, solitary and terminal. **Fruits** Obovate, flattened, greenish-brown achenes, 3-4 mm long. **Flowering** Winter, spring and summer. **Habitat** Widespread in well-drained soils along the coast, nearby ranges and inland slopes of southeastern Qld, eastern NSW, Vic. southeastern SA and Tas. **Family** Asteraceae.

Brachycome iberidifolia
Swan River Daisy

Wiry, branching, hairy annual herb to 50 cm high. **Leaves** Up to 4 cm long, finely divided into 5-13 narrow-linear segments. **Flowerheads** White, blue or violet, daisy-like, composite, solitary, terminal, 25-50 mm across with a yellow centre. **Fruits** Achenes. **Flowering** Winter, spring and summer. **Habitat** Sandhills, plains and watercourses of SA, the western half of WA and south central NT. **Family** Asteraceae.

Brachycome multifida
Cut-leaved Daisy

Spreading annual or perennial herb to 45 cm high. **Leaves** Narrow or broad-linear, to 7 cm long, finely divided into 7-10 linear or oblanceolate lobes to 35 mm long and about 1 mm wide. **Flowerheads** Mauve, blue, pink or white, daisy-like, composite, 20-25 mm across with a yellow centre, solitary and terminal on a stalk 4-14 cm long. **Fruits** Dark-brown to black, wedge-shaped, flattened achenes, 2-3 mm long. **Flowering** Autumn and winter. **Habitat** Widespread in open forests and grasslands of southern Qld, NSW and Vic. **Family** Asteraceae.

Brachycome rigidula
Hairy Cut-leaved Daisy

Perennial herb to 40 cm high with hairy stems. **Leaves** Finely divided, up to 2 cm long, with sharply-pointed segments. **Flowerheads** Blue, white or pink, composite, daisy-like, terminal, 1-3 cm across with a yellow centre. **Flowering** Winter, spring and summer. **Habitat** Rocky, well-drained soils above 500 m in southern Qld, NSW, Vic. and Tas. **Family** Asteraceae.

Minuria leptophylla
Minnie Daisy

Perennial herb to 50 cm high with erect, sparsely hairy branches. **Leaves** Alternate, narrow-linear with small points and minute hairs, 8-40 mm long and about 1 mm wide. **Flowerheads** White to purple with a yellow centre, composite, daisy-like, solitary and terminal, 15-30 mm across with 20-30 rays. **Fruits** Obovate brown to orange achenes, 1-3 mm long. **Flowering** Year round. **Habitat** Low shrubland, sclerophyll forests and woodlands inland in central western and southern Qld, the north coast and inland NSW, western Vic., SA, WA and southern NT. **Family** Asteraceae.

Brachycome aculeata

Calotis cuneifolia

Brachycome iberidifolia

Minuria leptophylla

Brachycome rigidula

Brachycome multifida

Senecio amygdalifolius
Peach-leaf Groundsel
Erect perennial herb or shrub to 150 cm high. **Leaves** Narrow-ovate to narrow-elliptic with regularly toothed margins, 5-12 cm long and 1-3 cm wide. **Flowerheads** Yellow, composite, daisy-like, about 2 cm across, with 4-8 rays. They are arranged in terminal clusters. **Fruits** Achenes, 4-6 mm long. **Flowering** Spring and summer. **Habitat** Tall forests and rainforest regrowth areas, mostly on coastal ranges of southeastern Qld, northern and central NSW. **Family** Asteraceae.

Senecio gregorii
Fleshy Groundsel
Erect annual herb to 40 cm high. **Leaves** Fleshy, broad-linear, 2-9 cm long and 2-5 mm wide. **Flowerheads** Brilliant yellow, terminal, solitary or in groups of 2-3 on leafy stalks 6-15 cm long, composite, daisy-like, 2-4 cm across, with 8-12 rays. **Fruits** Hairy, compressed, cylindrical, green or fawn achenes. **Flowering** Most of the year. **Habitat** Widespread in arid and semi-arid regions of western Qld, western NSW, northwestern Vic., SA, WA and central NT. **Family** Asteraceae.

Senecio magnificus
Showy Groundsel
Erect perennial herb or shrub to 1.5 m high. **Leaves** Alternate, crowded, often stem-clasping, fleshy, oblong to spathulate, blue-green, usually with coarsely-toothed margins, 3-9 cm long and 8-30 mm wide. **Flowerheads** Yellow, in terminal clusters of 10-50 on long, branching stems, composite, daisy-like, 3-4 cm across, with 4-8 large rays. **Fruits** Yellow to brown cylindrical achenes, 5-6 mm long. **Flowering** Winter and spring. **Habitat** Near streams and creek beds in dry inland areas of Qld, NSW, far western Vic., SA, WA and the NT. **Family** Asteraceae.

Senecio lautus
Variable Groundsel
Erect or sprawling perennial herb to 80 cm high. **Leaves** Variable, linear to lanceolate or spathulate, sometimes irregularly-toothed, deeply-lobed or pinnate, fleshy in sand dune sites, 1-8 cm long and 1-30 mm wide. **Flowerheads** Yellow, composite, daisy-like, 2-3 cm across, with 5-14 rays, arranged in terminal clusters of 10-25 flowerheads on stalks to 3 cm long. **Fruits** Brown or green cylindrical achenes 2-3 mm long, usually hairy. **Flowering** Most of the year. **Habitat** Widespread in a variety of habitats from the coast to inland in all states. **Family** Asteraceae.

Senecio linearifolius
Fireweed Groundsel
Erect, aromatic, perennial herb or soft shrub to 1.5 m high. **Leaves** Variable, linear to lanceolate or narrow-elliptic, usually toothed, 5-15 cm long and 5-30 mm wide, dark-green above and whitish-hairy below. **Flowerheads** Yellow, composite, daisy-like, 15-20 mm across, with 4-8 rays. They are arranged in large terminal clusters. **Fruits** Brown, compressed achenes, 2-3 mm long. **Flowering** Most of the year. **Habitat** Clay soils on sheltered slopes, mostly in wet sclerophyll forests of the coastal ranges in NSW, Vic. and Tas. **Family** Asteraceae.

Podolepis jaceoides
Showy Podolepis. Showy Copper Wire Daisy
Upright, perennial herb to 80 cm high with woolly stems. **Leaves** Alternate, oblanceolate, pointed, to 20 cm long and 2 cm wide, arranged in a rosette around the base of the plant. Stem leaves are narrow-lanceolate, 1-5 cm long and 2-10 mm wide. **Flowerheads** Yellow, composite, daisy-like, 2-3 cm across with 30-40 rays, solitary or 2-3 together, widely spaced and terminal, on stems 2-9 cm long with leaf-like bracts. **Fruits** Achenes about 3 mm long. **Flowering** Summer. **Habitat** Widespread from alpine areas to the western plains in woodlands and grasslands in all states except WA and the NT. **Family** Asteraceae.

Podolepis
jaceoides

Senecio
amygdalifolius

Senecio magnificus

Senecio
gregorii

Senecio lautus

Senecio linearifolius

Orites lancifolia
Alpine Orites

Upright or prostrate bushy shrub to 2 m high and 2 m diameter. **Leaves** Crowded, oblong to lanceolate, stiff and leathery, 1-3 cm long and 3-8 mm wide, glossy-green above and paler below. **Flowers** Cream-coloured, 4-5 mm long, with 4 curled-back petals and protruding stamens. They are arranged in crowded, terminal, cylindrical spikes, 2-5 cm long. **Fruits** Leathery follicles about 2 cm long. **Flowering** Summer. **Habitat** Common on rocky sites in alpine heaths and woodlands above 1200 m in the southern tablelands of NSW and eastern Vic. **Family** Proteaceae.

Eleocharis sphacelata
Tall Spike Rush

Erect, aquatic, perennial, leafless herb with a thick, hollow, flowering stem to 5 m high in deep water. **Flowers** Small, white and bristly, arranged in a cylindrical, solitary, terminal spike, 3-6 cm long and 8-9 mm wide. **Fruits** Bristly nuts, brown or straw-coloured, angled, obovoid to globular, 2-3 mm long. **Flowering** Spring, summer and autumn. **Habitat** Widespread in still fresh water to 5 m deep in ponds, swamps and waterways in all states except the NT. **Family** Cyperaceae.

Triglochin procera
Water Ribbons

Erect, robust, perennial aquatic herb to 80 cm high. **Leaves** Long and ribbon-like, semi-succulent, arising from and sheathing the base of the stem, to 2 m long and 4-40 mm wide, often floating on the water surface. **Flowers** Green with lilac styles, ovoid to urn-shaped, ridged, with 4-6 segments, to 5 mm across. They are arranged in dense, terminal, cylindrical spikes, 10-30 cm long and 3 cm wide, on a hollow stem 1-2 cm across. **Fruits** Globular follicles, 5-10 mm long. **Flowering** Spring and summer. **Habitat** Widespread in shallow water around ponds and streams in all states. **Family** Juncaginaceae.

Richea dracophylla
Dragon Heath

Erect, slightly branched shrub, usually 2 m, but sometimes up to 5 m high. **Leaves** Crowded at the ends of the branches, their bases sheathing the stem, spirally arranged, 15-30 cm long, narrow and tapering to a long, sharp point. **Flowers** White or pink, small and cup-shaped, to 12 mm long with long, protruding stamens. They are arranged in dense, terminal, hairy, cylindrical spikes, 12-25 cm long with large, brown, leaf-like bracts protruding from the tip. **Fruits** Tiny, segmented capsules, splitting open when dry. **Flowering** Spring. **Habitat** Mossy sites in montane sclerophyll forests and rainforests of Tas. **Family** Epacridaceae.

Lachnostachys verbascifolia
Lamb's Tails

Erect shrub to 1.2 m high, covered with soft hairs. **Leaves** Opposite, broadly oblong to ovate or lanceolate with curled-under margins, 3-8 cm long, covered with thick, white to brownish woolly hairs. **Flowerheads** White, woolly, axillary, cylindrical spikes, 6-10 cm long and 5-15 mm wide, with tiny, violet, tubular flowers set deeply into the woolly hairs. **Flowering** Winter and spring. **Habitat** Dry sand-heath plains in central western and southwestern WA. **Family** Verbenaceae.

Eleocharis sphacelata

Triglochin procera

Lachnostachys verbascifolia

Richea dracophylla

Orites lancifolia

Ptilotus exaltatus
Pink or Tall Mulla Mulla
Upright annual or perennial herb to 1.5 m high. **Leaves** Obovate or oblong-lanceolate to spathulate, to 20 cm long and 7 cm wide, thick, forming a rosette around the base of the plant. **Flowers** Lilac-grey to deep-pink, tubular, to 2 cm long, hairy with 5 small lobes, arranged in dense, terminal, conical to cylindrical spikes, 3-20 cm long and 30-45 mm wide. **Flowering** Most of the year. **Habitat** Widely distributed in arid areas of all mainland states. **Family** Amaranthaceae.

Ptilotus obovatus
Silver Mulla Mulla. Cotton Bush. White Foxtail
Compact, stiff, erect, perennial shrub to 1.2 m high, covered in white woolly hair. **Leaves** Elliptical to egg-shaped or lanceolate, covered with dense grey hairs, 1-6 cm long and 5-20 mm wide. **Flowers** White and pink to grey, tubular and hairy with 5 small lobes, 7-10 mm long, arranged in dense, terminal, short-cylindrical spikes, 1-3 cm long and about 15 mm wide. **Flowering** Most of the year. **Habitat** Widely distributed in arid areas of all mainland states. **Family** Amaranthaceae.

Ptilotus rotundifolius
Round Leaf Mulla Mulla
Erect, woolly shrub to 1.2 m high. **Leaves** Alternate, elliptical to orbicular with wavy margins, covered with soft grey hairs, 2-5 cm wide. **Flowers** Bright-pink, tubular and hairy with 5 small lobes, arranged in dense, woolly, terminal, cylindrical spikes to 12 cm long. **Flowering** Late winter and spring. **Habitat** Arid rocky hillsides in central western WA. **Family** Amaranthaceae.

Xanthorrhoea australis
Austral Grass Tree. Black Boy
Erect, very slow-growing shrub with a flowering stem to 5 m high and a thick trunk, sometimes to several metres high and branching. **Leaves** Blue-grey, long, stiff, linear, quadrangular in section, often more than 1 m long and 1-4 mm wide. **Flowerhead** is a creamy-white, velvety, cylindrical spike, 0.5-3.5 m long and 5-8 cm across, on a shorter woody stalk, packed with spirally-arranged clusters of small, 6-lobed flowers with 6 stamens, surrounded by hairless bracts. **Fruits** Shiny-brown, beaked capsules, 14-20 mm long. **Flowering** Late winter and spring, and often after fire. **Habitat** Widespread on poor sandy soils in heaths and open forests of NSW, southeastern Qld, Vic., southeastern SA and Tas. **Family** Xanthorrhoeaceae.

Xanthorrhoea minor
Small Grass Tree. Black Boy
Erect shrub with a buried trunk and a number of flowering stems, growing very slowly. **Leaves** Green, long, stiff, linear, triangular to concave in section, up to 60 cm long and 5 mm wide, tufted at the base of the plant. **Flowerhead** is a white to cream cylindrical spike, 5-12 cm long and 1-2 cm across, on a woody stalk 30-60 cm long and 3-8 mm across, packed with small, 6-lobed flowers with 6 protruding stamens, surrounded by hairy-fringed bracts. **Fruits** Shiny-brown, curved, beaked capsules, 14-16 mm long. **Flowering** Spring and often after fire. **Habitat** Heaths and swamps on poorly drained clay soils along the coast and tablelands of central eastern NSW, southeastern Qld, southwestern SA and Vic. **Family** Xanthorrhoeaceae.

Xanthorrhoea resinosa
Grass Tree. Black Boy
Erect shrub with a trunk less than 60 cm high or buried, and a flowering stem 0.4-2.5 m high and 8-30 mm diameter, growing very slowly. **Leaves** Long, stiff, linear, convex-triangular to quadrangular in section, 2-6 mm wide, tufted at the base of the plant. **Flowerhead** is a whitish, velvety, cylindrical spike, 60-220 cm long and 12-45 mm diameter, on a woody stalk, packed with small 6-lobed flowers with 6 protruding stamens, surrounded by densely-hairy, dark-brown bracts. **Fruits** Shiny-brown, beaked capsules. **Flowering** Spring and often after fire **Habitat** Widespread on wet sandy soils and swamps in heaths and low woodlands of the coast and tablelands o central eastern and southeastern NSW, southeastern Qld, and southeastern Vic. **Family** Xanthorrhoeaceae.

Ptilotus exaltatus

Xanthorrhoea minor

Ptilotus obovatus

Xanthorrhoea resinosa

Ptilotus rotundifolius

Xanthorrhoea australis

Callistemon brachyandrus
Prickly Bottlebrush

Erect, spreading shrub or small tree to 5 m high. **Leaves** Alternate, stiff, needle-like, sharply-pointed, finely-grooved above, 15-40 mm long and 1-2 mm wide. **Flowers** Orange-red with 5 small, green lobes and numerous protruding, crimson stamens with yellow tips, about 1 cm long, arranged in dense, hairy, cylindrical spikes, 3-5 cm long and 20-27 mm across with leafy shoots growing from the tip. **Fruits** Woody, cup-shaped capsules, 4-8 mm across, in cylindrical clusters around the branches. **Flowering** Spring. **Habitat** Scattered around creek beds in open woodlands, inland in NSW, northwestern Vic. and eastern SA. **Family** Myrtaceae.

Callistemon citrinus
Crimson Bottlebrush

Erect, rigid shrub to 3 m high. **Leaves** Alternate, lanceolate to narrow-elliptic, sharply-pointed, stiff, 3-7 cm long and 4-12 mm wide. **Flowers** Crimson with 5 small lobes and numerous long, protruding stamens, arranged in dense, hairy, cylindrical spikes, 5-12 cm long and 4-7 cm across with leafy shoots growing from the tip. **Fruits** Woody, cup-shaped capsules 4-7 mm across, in cylindrical clusters around the branches. **Flowering** Spring and summer. **Habitat** Widespread on damp, sandy flats and swamps near the coast and ranges of southeastern Qld, NSW and southeastern Vic. **Family** Myrtaceae.

Callistemon linearis
Narrow-leaved Bottlebrush

Erect shrub to 4 m high with silky young foliage. **Leaves** Alternate, rigid, channelled above, linear, sharply-pointed, 4-12 cm long and 1-3 mm wide. **Flowers** Red to dark crimson with 5 small lobes and numerous protruding stamens, 10-25 mm long. They are arranged in dense, hairy, cylindrical spikes, 5-10 cm long and 4-6 cm across with leafy shoots growing from the tip. **Fruits** Woody, cup-shaped capsules, 6-10 mm across, in cylindrical clusters around the branches. **Flowering** Spring and early summer. **Habitat** Widespread in damp places along the coast and ranges of central Qld and NSW. **Family** Myrtaceae.

Callistemon pachyphyllus
Wallum Bottlebrush

Dense, straggling shrub to 1.5 m high. **Leaves** Alternate, linear to lanceolate, 4-9 cm long and 3-15 mm wide, broader towards the tip, thick and leathery, flat with thickened margins. **Flowers** Crimson, rarely yellowish-green, with 5 small lobes and numerous protruding stamens about 25 mm long. They are arranged in dense, hairy, cylindrical spikes, 5-10 cm long and 4-6 cm across, with leafy shoots growing from the tip. **Fruits** Woody, grey-green, cylindrical capsules, 6-7 mm across, in cylindrical clusters around the branches. **Flowering** Summer. **Habitat** Swampy or sandy soils in coastal heaths in southeastern Qld and northeastern NSW. **Family** Myrtaceae.

Grevillea petrophiloides
Pink Pokers

Erect, rounded shrub to 2.5 m high. **Leaves** Much-divided into needle-like segments 2-5 cm long. **Flowers** Pink to red, sticky, tubular with 4 curled-back lobes and a long, wiry, protruding style when released, arranged in dense, cylindrical spikes, 5-10 cm long, on long, leafless, woody stems above the foliage. **Fruits** Shell-like sticky follicles 5-6 mm long. **Flowering** Winter and spring. **Habitat** Sandy or gravelly soils of the central coast and inland in WA. **Family** Proteaceae.

Melaleuca lateritia
Robin Red Breast Bush

Erect shrub to 2 m high with sparse lower foliage. **Leaves** Alternate, narrow-linear to lanceolate, often concave, about 12 mm long. **Flowers** Orange-red to scarlet, with 5 small lobes and numerous protruding stamens in bundles of 7-11, arranged in dense oblong to cylindrical hairy spikes, 5-7 cm long, below the ends of the branches. **Fruits** Woody, stalkless capsules, clustered along the branches. **Flowering** Summer. **Habitat** Thickets in clay soils of southwestern WA. **Family** Myrtaceae.

Grevillea petrophiloides

Melaleuca lateritia

Callistemon linearis

Callistemon brachyandrus

Callistemon pachyphyllus

Callistemon citrinus

Melaleuca decussata   **Totem Poles**

Erect shrub to 5 m high.**Leaves** Opposite, decussate, crowded, concave, linear to spathulate, sometimes pointed, 5-18 mm long and 1-3 mm wide, bluish-green and dotted with glands below. **Flowers** Pale-lilac with 5 small lobes and numerous protruding stamens, arranged in dense, cylindrical, hairy spikes about 2 cm long. **Fruits** Woody capsules, densely clustered along and embedded in the branches. **Flowering** Spring and summer. **Habitat** Rocky outcrops and sandy flats near streams in central to western Vic. and SA. **Family** Myrtaceae.

Melaleuca elliptica   **Oval Leaf or Granite Honey Myrtle**

Erect, bushy shrub to 3 m high. **Leaves** Opposite, oval, grey-green, to 12 mm long, dotted with glands below. **Flowers** Brilliant-scarlet with 5 small lobes and numerous protruding stamens, arranged in dense, oblong to cylindrical hairy spikes, 5-8 cm long and about 4 cm diameter, below the growing shoots. **Fruits** Woody, stalkless capsules, irregularly clustered along the branches. **Flowering** Spring and summer. **Habitat** Sandy soils among granite rocks in wet areas inland in southwestern WA. **Family** Myrtaceae.

Melaleuca wilsonii   **Violet or Wilson's Honey Myrtle**

Dense shrub with numerous slender branches to 2 m high. **Leaves** Opposite, decussate, concave, linear to lanceolate, 8-15 mm long and 1-2 mm wide, stalkless with sharp points. **Flowers** Pink to purple with 5 small lobes and numerous protruding stamens, arranged in dense, cylindrical or globular hairy spikes, 2-4 cm long, below the growing shoots. **Fruits** Stalkless woody capsules, irregularly clustered along the branches. **Flowering** Spring. **Habitat** Mallee scrubs in Vic. and SA. **Family** Myrtaceae.

Callistemon pallidus   **Lemon Bottlebrush**

Stiff, erect shrub to 8 m high. **Leaves** Alternate, narrow-elliptical, pointed, 25-70 mm long and 6-20 mm wide, stiff, leathery, dotted with glands. **Flowers** Yellow to cream with 5 small lobes and numerous long, protruding stamens, arranged in dense, hairy, cylindrical spikes, 3-9 cm long and 25-50 mm across with leafy shoots growing from the tip. **Fruits** Woody, cup-shaped capsules, 4-6 mm across, in cylindrical clusters around the branches. **Flowering** Spring and summer. **Habitat** Rocky sites near watercourses in the eastern ranges of southeastern Qld, NSW, Vic. and Tas. **Family** Myrtaceae.

Callistemon rugulosus (syn. C. macropunctatus)   **Scarlet Bottlebrush**

Straggling, prickly shrub to 4 m high. **Leaves** Alternate, linear to narrow-lanceolate, sharply-pointed, 2-6 cm long and 3-7 mm wide, stiff, dotted with glands. **Flowers** Bright-red with 5 small lobes and numerous long, yellow-tipped, protruding stamens, arranged in dense, hairy, cylindrical spikes, 5-10 cm long and 4-5 cm across, with leafy shoots growing from the tips. **Fruits** Woody, cup-shaped capsules in cylindrical clusters around the branches. **Flowering** Spring. **Habitat** Sandy or swampy sites in mallee scrubs and low, open forests in western Vic. and SA. **Family** Myrtaceae.

Callistemon sieberi   **Alpine or River Bottlebrush**

Prickly, rounded shrub to 5 m high. **Leaves** Alternate, linear to narrow-lanceolate, sharply-pointed, erect, 1-5 cm long and 1-5 mm wide, thick with scattered glands below. **Flowers** Golden-yellow with 5 small lobes and numerous long, protruding stamens, arranged in dense, hairy, cylindrical spikes, 2-6 cm long and 20-25 mm across, with leafy shoots growing from the tips. **Fruits** Woody, cup-shaped capsules 4-5 mm across in cylindrical clusters around the branches. **Flowering** Spring and summer. **Habitat** Wet sites in heaths and woodlands in the ranges of southeastern Qld, NSW, Vic., southeastern SA and northeastern Tas. **Family** Myrtaceae.

Callistemon rugulosus

Callistemon sieberi

Melaleuca elliptica

Melaleuca decussata

Callistemon pallidus

Melaleuca wilsonii

Banksia canei
Mountain Banksia

Dense, prickly, flat-topped shrub to 3 m high and 4 m wide. **Leaves** Alternate, stiff, narrow-elliptic to obovate, often with irregular spines on curved-back margins, 2-8 cm long and 4-10 mm wide, shiny-green above and whitish below with rusty hairs. **Flowers** Yellowish-green, tubular, 18-20 mm long with long, straight, wiry, protruding styles, arranged in terminal, cylindrical spikes, 4-15 cm long and 2-6 cm diameter. **Fruits** Brown or grey cones, with up to 150 follicles, 12-18 mm long, hairy, remaining closed until burnt. **Flowering** Summer and autumn. **Habitat** Low woodlands and heaths on rocky slopes and gullies between 750 and 1500 m altitude in the southern tablelands of NSW and eastern Vic. **Family** Proteaceae.

Banksia baueri
Possum or Woolly-spiked Banksia

Dense, rounded shrub to 5 m high. **Leaves** Rigid, oblong to spathulate, 7-15 cm long and 1-2 cm wide, with evenly-toothed, prickly margins and prominent veins. **Flowers** Small and tubular, lemon-yellow with grey-mauve, long, wiry, straight, protruding styles, arranged in large cylindrical spikes, 15-40 cm long and 12-20 cm diameter. **Fruits** Cones with furry follicles. **Flowering** Winter and spring. **Habitat** Open sand heaths of the coast and inland areas around the Stirling, Avon and Eyre districts of WA. **Family** Proteaceae.

Melaleuca gibbosa
Slender Honey Myrtle

Dense, wiry shrub to 3 m high. **Leaves** Opposite, decussate, crowded, concave, ovate, 2-7 mm long and 1-4 mm wide, thick and rigid, pale to grey-green. **Flowers** Pink-mauve with 5 small lobes and 15-20 long, protruding stamens in bundles, arranged in dense, cylindrical to globular, hairy, leafy spikes, about 15 mm long with new shoots extending from the tip. **Fruits** Stalkless woody capsules about 5 mm across, densely clustered and embedded in the branches. **Flowering** Spring and summer. **Habitat** Damp areas in heaths and open forests in southwestern Vic., SA, southeastern WA and Tas. **Family** Myrtaceae.

Verticordia grandis
Scarlet Feather Flower

Upright or spreading shrub with many rigid branches, to 2 m high. **Leaves** Opposite, orbicular, blue-grey, 8-12 mm diameter. **Flowers** Scarlet, tubular, to 25 mm across with 5 much-divided, feathery lobes and a long, protruding style, arranged in crowded cylindrical spikes among the upper leaves. **Flowering** Most of the year, but mainly in summer. **Habitat** Open sand heaths between Perth and Geraldton in WA. **Family** Myrtaceae.

Calothamnus villosus
Silky or Hairy Net Bush

Spreading shrub with hairy branches to 3 m high. **Leaves** Crowded, needle-like or slightly flattened, sometimes sharply-pointed, grey-green, covered with scattered hairs, 1-3 cm long. **Flowers** Red, tubular, with 5 small lobes and 5 long, feathery, protruding stamens, arranged in dense spikes about 10 cm long, usually on one side of the branch. **Fruits** Globular capsules about 1 cm diameter. **Flowering** Spring and summer. **Habitat** Gravelly soils in open forests of southwestern WA. **Family** Myrtaceae.

Beaufortia sparsa
Swamp or Gravel Bottlebrush

Erect, spreading shrub to 4 m high. **Leaves** Crowded but scattered along the branches, ovate to elliptical, sometimes with curved-back margins, 7-10 mm long and 3-5 mm wide. **Flowers** Bright-scarlet, rarely white with 5 small orbicular lobes and very long, filamented, protruding stamens. They are arranged in dense cylindrical to globular spikes 6-7 cm across, below the new growth. **Fruits** Stalkless, woody, cylindrical capsules, about 5 mm long, clustered along the branches. **Flowering** Summer. **Habitat** Sandy and peaty swamps of southwestern WA. **Family** Myrtaceae.

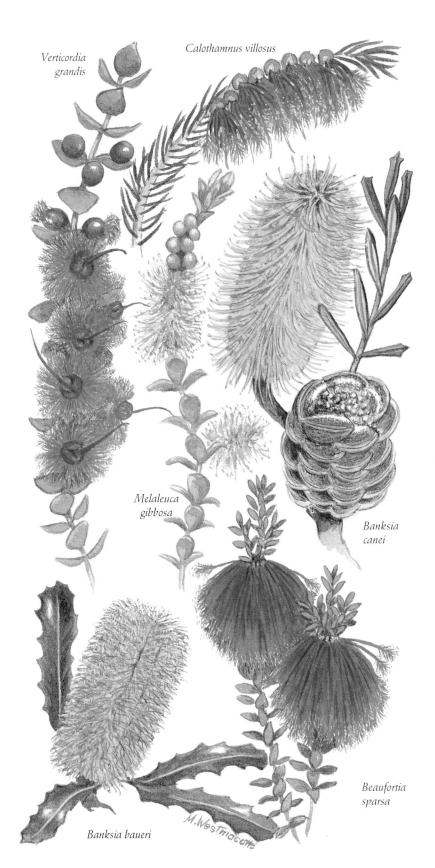

Verticordia grandis

Calothamnus villosus

Melaleuca gibbosa

Banksia canei

Banksia baueri

Beaufortia sparsa

M.Westmacott

Banksia aspleniifolia
Fern-leaved Banksia

Low, straggly or compact shrub to 2 m high with rusty hairs on young shoots. **Leaves** Leathery, irregularly toothed, elliptical-oblong to obovate, broader towards the tip, 5-12 cm long and 15-25 mm wide, white and downy below. **Flowers** Lemon-yellow, small and tubular with long, wiry, straight styles, arranged in terminal, cylindrical spikes 6-15 cm long and 4-7 cm diameter. **Fruits** Oblong cones, retaining the seeds until burnt. **Flowering** Autumn and early winter. **Habitat** Sandy and rocky soils in southeastern Qld. **Family** Proteaceae.

Banksia collina (syn. B. spinulosa)
Hill Banksia

Stiff shrub with smooth, grey stems to 4 m high, often regarded as a variety of *B. spinulosa*. **Leaves** Broad-linear to narrow-oblong with evenly toothed, often curved-back margins, 2-12 cm long and 2-8 mm wide, dark-green above and whitish below. **Flowers** Yellow to bronze, tubular, 2-3 cm long with long, hooked, wiry, protruding styles, black with yellow tips, arranged in cylindrical spikes along the branches, 5-20 cm long and 4-9 cm diameter. **Fruits** Narrow cones, 10-24 mm long, crowded, with up to 100 follicles. **Flowering** Autumn, winter and early spring. **Habitat** Dry sclerophyll forests, woodlands and heaths of the coast and ranges of southeastern Qld, central eastern and northeastern NSW. **Family** Proteaceae.

Banksia elderana
Palm or Swordfish Banksia

Dense, multi-stemmed shrub to 4 m high with hairy branches. **Leaves** Long and narrow, 15-60 cm long and 1-2 cm wide, with evenly toothed, prickly margins. **Flowers** Small and tubular, bright-yellow with long, wiry, straight, protruding styles, arranged in pendant, ovoid or oblong cylindrical spikes, 10-15 cm long and 6-10 cm diameter. **Fruits** Brown to grey cones, 8-12 cm long and 6-8 cm across. **Flowering** Winter, spring and summer. **Habitat** Sandy, open country in central southwestern WA. **Family** Proteaceae.

Banksia ornata
Desert Banksia

Dense, rounded shrub to 2.5 m high. **Leaves** Greyish-green, hairy, stiff, oblong to spathulate, 5-10 cm long and 2-3 cm wide, with evenly-toothed margins. **Flowers** Small and tubular, yellow to bronze with grey, long, wiry, straight, protruding styles, arranged in ovoid or oblong cylindrical spikes, 5-14 cm long. **Fruits** Grey-brown cones retaining the seeds until burnt. **Flowering** Autumn and winter. **Habitat** Sandy coastal and inland areas of southeastern SA and southwestern Vic. **Family** Proteaceae.

Banksia robur
Swamp Banksia

Erect, spreading shrub to 2 m high with dense, rusty hairs on the branchlets. **Leaves** Alternate, leathery, large, obovate to elliptic, 10-30 cm long and 5-10 cm wide, with irregularly toothed, spiny margins, glossy dark-green above and covered in white or rusty hairs below. **Flowers** Tubular, 20-25 mm long, bluish to yellow-green with long, wiry, straight, protruding styles tipped with black, arranged in terminal, cylindrical spikes, 6-17 cm long and 7-8 cm diameter. **Fruits** Hairy cones, 10-16 mm long, with up to 100 follicles, opening after fire. **Flowering** From summer to winter. **Habitat** Heaths and woodlands in sandy, swampy areas along the east coast of Qld and the northern and central coast of NSW. **Family** Proteaceae.

Banksia spinulosa
Hairpin Banksia

Erect, straggling, multi-stemmed shrub to 3 m high. **Leaves** Alternate, scattered, narrow-linear to oblong or narrow-elliptic, 2-10 cm long and 1-10 mm wide, whitish below, notched at the ends with a prominent point and several small teeth, sometimes with toothed margins. **Flowers** Tubular, 2-3 cm long, bronze or honey-coloured with black, protruding, wiry, hooked styles, arranged in cylindrical spikes, 5-20 cm long and 4-8 cm diameter. **Fruits** Narrow cones, 10-24 mm long, with up to 100 crowded follicles. **Flowering** Autumn and winter. **Habitat** Widespread in heaths and open forests of the coast and ranges of Qld, NSW and eastern Vic. **Family** Proteaceae.

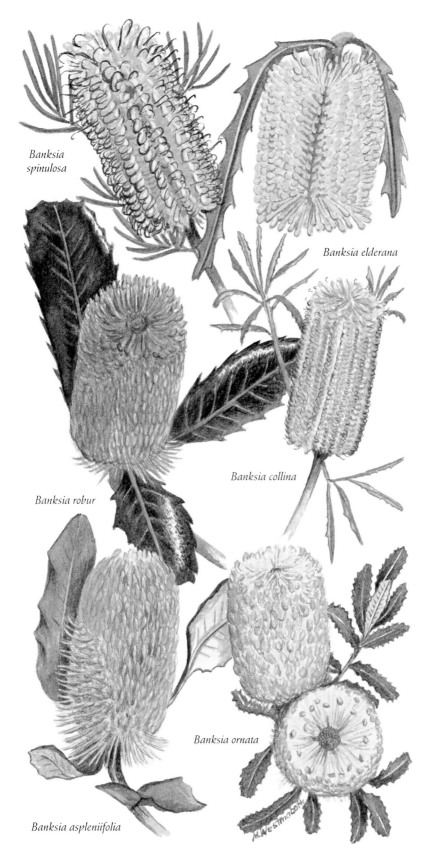

Banksia spinulosa

Banksia elderana

Banksia robur

Banksia collina

Banksia aspleniifolia

Banksia ornata

M. Westmacott

Banksia ashbyi
Orange Banksia

Erect shrub to 4 m high. **Leaves** 15-20 cm long and about 3 cm wide, flat, smooth, deeply cut into numerous tri-angular lobes. **Flowers** Rich-orange, small and tubular with long, wiry, straight styles, arranged in ovoid to cylindrical terminal spikes, 10-15 cm long and 7-10 cm diameter. **Flowering** Winter and spring. **Habitat** Sandy soils in open woodlands and heaths of the central coast of WA. **Family** Proteaceae.

Banksia burdettii

Stiff, spreading shrub to 6 m high. **Leaves** Rigid, narrow-elliptic or oblong, 7-25 cm long and about 2 cm wide, sharply-toothed, pale-green below, often with woolly hairs. **Flowers** Orange, small and tubular with long, straight, wiry, protruding styles, arranged in terminal, acorn-shaped to oblong, cylindrical spikes, 7-12 cm long and 8-12 cm diameter. **Fruits** Oblong cones retaining the seeds. **Flowering** Late spring and summer. **Habitat** Open woodlands and sandy heaths of southwestern WA between Moore and Hill rivers. **Family** Proteaceae.

Banksia coccinea
Scarlet Banksia

Multi-stemmed, erect shrub to 5 m high with furry branches. **Leaves** Leathery, broad-oblong or oval, 5-10 cm long and 4-8 cm wide, with irregular spiny margins, light-green above and white or greyish below. **Flowers** Small and tubular, grey with bright-scarlet, wiry, straight, protruding styles, tipped with gold. They are arranged in vertical rows on a terminal, squat, cylindrical spike, 6-12 cm long and 6-15 cm diameter, set in a rosette of leaves. **Fruits** Cones to about 10 cm long. **Flowering** Winter, spring and summer. **Habitat** Gravelly, sandy or marshy areas on the southwestern and southern coast of WA. **Family** Proteaceae.

Banksia hookerana
Acorn Banksia

Compact shrub to 3 m high with marbled grey and orange bark and hairy branches. **Leaves** Narrow-linear, 10-25 cm long and about 1 cm wide with evenly toothed margins. **Flowers** Small and tubular, velvety-white before opening, yellow-orange when open with long, straight, protruding styles, arranged in terminal, acorn-shaped cylindrical spikes, 8-10 cm long and 6-8 cm diameter. **Fruits** Cones about 10 cm long and 5 cm across. **Flowering** Winter, spring and summer. **Habitat** Open thickets in light, sandy soil of the central coast of WA. **Family** Proteaceae.

Banksia gardneri (syn. B. prostrata)
Prostrate Banksia

Prostrate shrub with spreading, hairy branches. **Leaves** Variable, erect, stiff, linear-oblong, 15-40 cm long and 2-8 cm wide with deeply toothed margins. **Flowers** Small and tubular, brownish-yellow with long, wiry, straight, protruding, yellow styles, arranged in terminal, erect, cylindrical spikes, 7-15 cm long and 6-8 cm diameter. **Fruits** Cones about 10 cm long. **Flowering** Spring and summer. **Habitat** Coastal sand heaths of southwestern WA. **Family** Proteaceae.

Banksia repens
Creeping Banksia

Prostrate shrub with erect leaves and flower spikes and hairy branches. **Leaves** Variable, 20-50 cm long and 4-15 cm wide, cut virtually to the prominent midrib, each segment being either lobed or entire. **Flowers** Small and tubular, yellowish brown to rose-red with long, wiry, protruding styles, curved inwards, with a yellow tip. They are arranged in cylindrical spikes, 10-15 cm long and 6-7 cm diameter. **Fruits** Cones, 10-15 cm long. **Flowering** Spring and early summer. **Habitat** Open sand heaths in the Stirling and Eyre districts of southern WA **Family** Proteaceae.

Banksia ashbyi

Banksia burdettii

Banksia hookerana

Banksia coccinea

Banksia repens

Banksia gardneri

Grevillea acanthifolia

Straggling or prostrate shrub to 3 m high. **Leaves** Stiff and prickly, 3-12 cm long and 3-7 cm wide, pinnate, divided almost to the midrib, with 2-15 linear to lanceolate or wedge-shaped, pointed lobes, 5-20 mm long and 2-3 mm wide. **Flowers** Pink to purple and white, hairy outside, small and tubular, about 8 mm long with wiry, protruding styles about 2 cm long, red with green tips, initially hooked, but erect when released. They are arranged in terminal or axillary, spidery or comb-like, one-sided spikes, 3-10 cm long. **Fruits** Hairy, thin-walled follicles with reddish-brown stripes, containing 2 seeds. **Flowering** Spring and summer. **Habitat** Near swamps and watercourses along the north coast and the ranges of eastern NSW. **Family** Proteaceae.

Grevillea bipinnatifida

Low, sprawling or prostrate shrub, to 1.5 m high, with hairy branches. **Leaves** Bipinnate, rigid and prickly, to 20 cm long, divided into 9-21 oblong or wedge-shaped pointed segments which are again divided. **Flowers** Red, curled tubes to 16 mm long, silky-hairy outside with long, curved, protruding styles when released. They are arranged in pendulous, one-sided, spidery-looking spikes to 20 cm long. **Fruits** Thin-walled follicles with 2 seeds. **Flowering** Winter and spring. **Habitat** Gravel and granite soils of the Darling Range in WA. **Family** Proteaceae.

Grevillea hookeriana   Red Toothbrush Grevillea

Spreading shrub to 3 m high with hairy branches. **Leaves** Divided into 3-9 needle-like wiry segments, 7-20 cm long, with curled-under margins and a prominent midrib. **Flowers** Bright-red, curled tubes, silky-hairy outside with long, curved, protruding styles when released. They are arranged in one-sided, spidery-looking spikes to 8 cm long. **Fruits** Shell-like follicles about 2 cm long. **Flowering** Most of the year. **Habitat** Granite rocks and sandy soils in southwestern WA. **Family** Proteaceae.

Grevillea juniperina   Prickly Grevillea

Rigid, rounded, erect or prostrate shrub to 3 m high with short, hairy branches. **Leaves** Linear to narrow-lance-olate or narrow-elliptic, rigid, sharply-pointed with rolled-under margins, silky-hairy below, 5-35 mm long and 1-6 mm wide. **Flowers** Bright-red to pink or cream, tubular with curled-back, satiny petals, about 15 mm long with long, protruding, curved styles when released. They are arranged in small, spidery-looking, terminal clusters, 25-35 mm long. **Fruits** Thin-walled follicles, 8-10 mm long. **Flowering** Mainly from winter to early summer. **Habitat** Wet areas, often by watercourses of the coast and ranges of eastern NSW. **Family** Proteaceae.

Grevillea oleoides   Olive Spider Flower

Erect, sparse shrub to 3 m high. **Leaves** Linear to narrow-obovate or elliptical with curved-back margins, 5-14 cm long and 4-20 mm wide, dark-green with silky hairs below. **Flowers** Deep-red tubes with curled-back petals, 8-14 mm long, silky-hairy with long, curved, protruding styles when released. They are arranged in small, axillary or terminal, spidery-looking clusters, 2-4 cm long. **Fruits** Thin-walled follicles about 12 cm long with 2 seeds. **Flowering** Mainly in winter and spring. **Habitat** Moist, sandy and rocky sites, often by creeks or in swampy areas, in heaths and dry sclerophyll woodlands of the coast and ranges of central eastern NSW. **Family** Proteaceae.

Grevillea speciosa (syn. G. punicea)   Red Spider Flower

Erect, spindly shrub to 3 m high with silky-hairy young branches. **Leaves** Obovate, elliptic or narrow-elliptic, 1-5 cm long and 4-12 mm wide with curved-back margins, shiny green above and whitish with a prominent midrib and silky hairs below. **Flowers** Bright-red or rarely white, tubular, 8-10 mm long with 4 curled-back petals and a curved, protruding style, 20-25 mm long when released. They are arranged in small, crowded, terminal, spidery-looking clusters, 2-4 cm long. **Fruits** Hairless, thin-walled follicles with 2 seeds. **Flowering** Most of the year. **Habitat** Moist sites on sandy soils in heaths and dry sclerophyll forests of the central coast and nearby ranges of NSW **Family** Proteaceae.

Grevillea bipinnatifida

Grevillea oleoides

Grevillea acanthifolia

Grevillea speciosa

Grevillea hookeriana

Grevillea juniperina

Grevillea aquifolium
Prickly Grevillea

Spreading or upright shrub to 2 m high. **Leaves** Stiff, dull-green above, paler and hairy below, oblong, 15-80 mm long and 5-27 mm wide, with sharp, spiny lobes. **Flowers** Green and hairy, tubular, 5-9 mm long with red, wiry styles, 2-3 cm long, initially hooked, but erect when released. They are arranged in terminal, spidery or comb-like, one-sided spikes. **Fruits** Thin-walled, compressed, ellipsoid follicles, 7-11 mm long, with 2 seeds. **Flowering** Spring and summer. **Habitat** Sandy soils in heaths and open country of western Vic and southeastern WA. **Family** Proteaceae.

Grevillea banksii

Tall, slender or spreading shrub to 7 m high, or sprawling and prostrate. **Leaves** Pinnate with 3-11, deeply divided, linear to lanceolate segments, 5-10 cm long and about 1 cm wide, with curved-back margins, smooth and dark-green above, covered with silky hairs below. **Flowers** Bright-red or creamy-white, slightly hairy, curled tubes, about 15 mm long with long, curved, protruding styles when released. They are arranged in terminal, spidery-looking spikes to 15 cm long. **Fruits** Furry brown, compressed, ovate follicles, about 25 mm long with 2 seeds. Flowering most of the year, but mainly in spring. **Habitat** Poor soils and sites in barren country along the east coast of Qld. **Family** Proteaceae.

Grevillea huegelii
Comb Grevillea

Rigid, prickly, spreading to prostrate shrub to 3 m high. **Leaves** Stiff, crowded, 15-60 mm long, divided into 2-11 (rarely 25) sharply-pointed, linear lobes, 4-40 mm long and 1-2 mm wide, with a double groove below. **Flowers** Deep-pink to red, tubular, 15-29 mm long with 4 curled-back petals and long, protruding, straight styles 10-12 mm long when released. They are arranged in crowded, spidery-looking clusters, 25-50 mm long. **Fruits** Thin-walled, hairless, compressed, ellipsoid follicles, 10-12 mm long with 2 seeds. **Flowering** Spring and summer. **Habitat** Mainly mallee scrubs and belah woodlands in dry inland areas of NSW, Vic., SA and WA. **Family** Proteaceae.

Grevillea ilicifolia
Holly Grevillea

Spreading or prostrate shrub to 2 m high with hairy branches. **Leaves** Stiff, prickly and holly-like, 2-11 cm long and 1-6 cm wide, usually with 3-10 obovate to wedge-shaped lobes, 5-50 mm long and 4-8 mm wide, with curled-back margins, whitish and densely hairy below. **Flowers** Pale-green, curled tubes, 5-10 mm long with red, protruding styles, 17-25 mm long when released. They are arranged in short, one-sided, sub-terminal spikes, 2-5 cm long. **Fruits** Thin-walled, hairy, ovoid follicles with reddish-brown blotches, 10-14 mm long with 2 seeds. **Flowering** Mainly in spring. **Habitat** Low mallee or heath on sandy soils in southwestern NSW, western Vic. and eastern SA. **Family** Proteaceae.

Grevillea longifolia
Fern-leaf Spider Flower

Spreading shrub or small tree to 5 m high. **Leaves** Alternate, stiff, linear to narrow-lanceolate, 10-25 cm long and 10-20 mm wide, with entire, coarsely-toothed or deeply-divided margins, covered with silky hairs below. **Flowers** Dark-red curled tubes, silky-hairy outside, 5-10 mm long with long, red, protruding styles, 20-14 mm long when released. They are arranged in short, one-sided, terminal or axillary spikes, 45-75 mm long. **Fruits** Thin-walled follicles with reddish-brown markings and 2 seeds. Flowering mainly in spring. **Habitat** Moist gullies and riverbanks in dry sclerophyll forests on sandstone along the NSW central coast and Blue Mountains. **Family** Proteaceae.

Grevillea sericea
Silky or Pink Spider Flower

Usually a low, erect, bushy shrub to 2 m high with silky hairs on leaves, flowers and young branches. **Leaves** Often in whorls of 3, elliptic to oblanceolate or linear, 1-9 cm long and 2-9 mm wide, with small pointed tips, a prominent midrib, curved-back margins and dense, silky hairs below. **Flowers** Deep-pink to white, tubular, bearded inside, 4-8 mm long with 4 curled-back hairy petals and a curved, protruding, pink style, 10-19 mm long when released. They are arranged in dense, short, terminal, spidery-looking clusters, 15-60 mm long. **Fruits** Boat-shaped, hairless follicles, 12-15 mm long and 5-7 mm wide. **Flowering** Most of the year. **Habitat** Sandy soils in dry sclerophyll forests of southeastern Qld, the central coast and nearby ranges of NSW. **Family** Proteaceae.

*Grevillea
banksii*

Grevillea huegelii

*Grevillea
aquifolium*

Grevillea sericea

Grevillea ilicifolia

*Grevillea
longifolia*

Grevillea arenaria

Erect, spreading shrub to 3 m high with hairy branches. **Leaves** Obovate to elliptic or linear, with a small point and curved-back margins, 15-75 mm long and 3-15 mm wide, dull-green above with silky hairs below. **Flowers** Red, pink or yellow-green, curled, hairy tubes, bearded inside, 8-10 mm long, with straight, wiry, protruding styles 24-33 mm long when released. They are arranged in terminal, spidery clusters of 2-10 flowers. **Fruits** Thin-walled, hairy follicles to 12 mm long with 2 seeds. **Flowering** Year round, particularly in spring. **Habitat** Sandy soils in open forests in rocky sites, usually near creeks or cliffs, on the coast, tablelands and western slopes of NSW. **Family** Proteaceae.

Grevillea lanigera   Woolly Grevillea

Greyish-green, rounded, hairy shrub to 2 m high. **Leaves** Crowded, narrow-oblong to lanceolate or linear, with rolled-under margins, 1-3 cm long and 1-5 mm wide, covered with soft hairs below. **Flowers** Red or pink and cream, curled tubes, bearded inside, about 1 cm long with, hairy, protruding styles, 13-20 mm long when released. They are arranged in small, terminal, compact, spidery clusters. **Fruits** Hairy, thin-walled follicles, 10-12 mm long with 2 seeds. **Flowering** Winter and spring. **Habitat** Moist sandy or rocky sites, on stream banks, coastal heaths and open forests, to the sub-alps along the central and south coast and ranges of NSW and eastern Vic. **Family** Proteaceae.

Grevillea lavandulacea   Lavender Grevillea

Low, spreading or prostrate shrub to 1 m high. **Leaves** Variable, rigid, normally linear to elliptic, oblong or lanceolate with curved-back margins and sharp points, 2-35 mm long and 1-8 mm wide, covered with dense pale hairs below. **Flowers** Red to pink and white, slightly hairy, curled tubes, 6-11 mm long with protruding styles 17-25 mm long when released. They are arranged small, terminal, semi-erect, spidery clusters. **Fruits** Slightly hairy ovoid follicles, 12-15 mm long and 7 mm wide. **Flowering** Winter and spring. **Habitat** Mallee scrub and low rainfall areas of western Vic. and eastern SA. **Family** Proteaceae.

Grevillea rosmarinifolia   Rosemary Grevillea

Dense or open shrub to 2 m high with slender, velvety branches. **Leaves** Narrow-linear to narrow-elliptic, stiff with sharp points and curled-under margins, 1-4 cm long and 1-3 mm wide. **Flowers** Red to creamy-pink, curled tubes, bearded inside, 6-8 mm long with curved, protruding styles 15-23 mm long when released. They are arranged in short, terminal, spidery clusters of 1-8 flowers. **Fruits** Hairy, thin-walled follicles with 2 seeds. **Flowering** Spring and summer. **Habitat** Moist sites near streams in open forests, and also in mallee communities on the sandy plains of Vic. and the central and southern tablelands and western slopes of NSW. **Family** Proteaceae.

Grevillea victoriae   Royal Grevillea

Rounded shrub to 4 m high with silky-hairy branches. **Leaves** Variable, narrow-lanceolate to broad-elliptic, 1-11 cm long and 5-45 mm wide, often with whitish hairs below. **Flowers** Red to rusty-red, tubular, silky outside and bearded within, 10-16 mm long with 4 curled-back, hairy petals, and protruding styles 16-26 mm long when released. They are arranged in small, pendulous, terminal clusters, 20-85 mm long. **Fruits** Thin-walled, hairless, narrow follicles to 2 cm long with 2 seeds. **Flowering** Spring and summer. **Habitat** Rocky sites on ridges, slopes and stream banks, in sclerophyll forests, heaths and snow gum woodlands above 1200 m in the southern tablelands of NSW and eastern Vic. **Family** Proteaceae.

Grevillea wilsonii

Spreading shrub to 1.5 m high. **Leaves** Divided into several spreading, needle-like, forking, prickly, rigid segments, 12-25 mm long, with a double groove below. **Flowers** Scarlet, curled tubes, bearded inside, 12-15 mm long with long, curved, protruding styles when released. They are arranged in loose clusters at the ends of the branchlets. **Fruits** Shell-like follicles. **Flowering** Spring. **Habitat** Gravelly sites in southwestern WA. **Family** Proteaceae.

Grevillea victoriae

Grevillea
rosmarinifolia

Grevillea wilsonii

Grevillea
lanigera

Grevillea
lavandulacea

Grevillea arenaria

Grevillea alpine
Cat's Claws

Low, spreading or semi-prostrate shrub to 2 m high, covered with downy hairs. **Leaves** Soft, crowded, grey-green, ovate or oblong to oblanceolate, 5-25 mm long and 1-10 mm wide with rolled-under margins, silky-hairy below. **Flowers** Yellow, orange and red to green and white, tubular with curled-back lobes, bearded inside, 6-10 mm long, with wiry, hairy styles, red with yellowish tips, 10-21 mm long, initially hooked but erect when released. They are arranged in loose, terminal, spidery clusters. **Fruits** Hairy follicles about 15 mm long with 2 seeds. **Flowering** From spring to early summer. **Habitat** Heaths, open forests and mallee, on sandy soils in Vic., the southern tablelands and south west slopes of NSW. **Family** Proteaceae.

Grevillea buxifolia
Grey Spider Flower

Erect to spreading bushy shrub to 2 m high, with dense, rusty hairs on the branches. **Leaves** Oblong, broad-lanceolate or ovate with curved-back margins, 6-35 mm long and 2-9 mm wide, shiny-green above and hairy below. **Flowers** Grey curled tubes, woolly outside and bearded inside, 4-6 mm long, with curved styles 10-17 mm long when released. They are arranged in dense, terminal, spidery clusters. **Fruits** Hairy, thin-walled follicles with 2 seeds. **Flowering** From spring to autumn. **Habitat** Sandstone areas in dry sclerophyll woodlands and heaths of the central and southern coast and adjacent plateaus in NSW. **Family** Proteaceae.

Grevillea chrysophaea
Golden Grevillea

Spreading shrub to 1.5 m high. **Leaves** Oval with a tiny, rigid point and curved-back margins, 15-35 mm long and 5-15 mm wide, smooth and green above, pale and felted below. **Flowers** Golden-brown or greenish, furry, curled tubes about 15 mm long with a long, hairy, protruding style when released. They are arranged in small terminal clusters. **Fruits** Thin-walled follicles, 10-15 mm long with 2 seeds. **Flowering** Spring. **Habitat** Stony or sandy soils in dry, open forests along the coast and hills of southeastern Vic. **Family** Proteaceae.

Grevillea floribunda
Seven Dwarfs Grevillea

Erect, straggling shrub to 2 m high. **Leaves** Lanceolate to elliptical with curved-back margins, 2-8 cm long and 2-20 mm wide, dull-green above, pale and finely-hairy below. **Flowers** Curled tubes, 7-10 mm long, yellowish to brown-green or orange with rusty-brown hairs outside, bearded inside, with straight, protruding styles, 1-2 cm long when released. They are arranged in terminal, pendulous, spidery clusters of 6-20 flowers. **Fruits** Hairy, thin-walled follicles, 10-15 mm long with 2 seeds. **Flowering** Mainly in spring. **Habitat** Sandy or gravelly soils in open forests and rocky ridges of the tablelands, western slopes and plains of NSW and southeastern Qld. **Family** Proteaceae.

Grevillea juncifolia
Honeysuckle Spider Flower

Erect straggling shrub or small tree to 7.5 m high with rough, grey bark. **Leaves** Linear with curled-under margins, rigid, pointed and covered with soft hair, 10-30 cm long and 1-2 mm wide, simple or divided into 2,3 or 6 lobes, 5-22 cm long and 1-2 mm wide. **Flowers** Orange-yellow curled tubes, 7-12 mm long, covered with soft hairs, with curved, protruding styles, 16-27 mm long when released. They are arranged in terminal, spidery-looking clusters, 7-20 cm long. **Fruits** Flat, ovate, hairy follicles, with reddish-brown markings, 20-25 mm long. Flowering year round, especially in winter and spring. **Habitat** Sandy soils and sand dunes in woodlands, inland in all mainland states except Vic. **Family** Proteaceae.

Grevillea linearifolia *(syn. G. parviflora)*
Small-flower Grevillea

Erect, rounded shrub to 3.5 m high. **Leaves** Widely-spreading, rigid, linear to narrow-elliptic, pointed, 15-100 mm long and 1-5 mm wide with curled-under margins, bright-green above, with a double groove and whitish hairs below. **Flowers** White or pale-pink, tubular with 4 curled-back petals, 3-5 mm long, minutely-hairy inside, with curved, protruding styles, 5-13 mm long when released. They are arranged in small, terminal, spidery clusters. **Fruits** Thin-walled, hairless, ellipsoid follicles about 12 mm long with 2 seeds. Flowering mainly in spring and summer. **Habitat** Widespread on sandy and rocky sites in heaths and dry sclerophyll forests of the coast and ranges of NSW, Vic. and SA. **Family** Proteaceae.

Grevillea alpine

Grevillea buxifolia

Grevillea juncifolia

Grevillea chrysophaea

Grevillea linearifolia

Grevillea floribunda

Grevillea candicans

Erect shrub to 2 m high. **Leaves** Alternate, 9-15 cm long, divided almost to the midrib into many linear lobes, 5-9 cm long and 1-2 mm wide. **Flowers** Creamy-white, curled tubes with long, curved styles when released, arranged in terminal racemes 5-7 cm long. **Flowering** Late winter and spring. **Fruits** Thin-walled follicles. **Habitat** Sandy, coastal heaths in southwestern and southern WA. **Family** Proteaceae.

Hakea nodosa   Yellow Hakea

Erect, bushy shrub to 4 m high with ribbed branches. **Leaves** Needle-like to narrow-linear, rather soft, 1-5 cm long and 1-3 mm wide. **Flowers** Yellow to cream and strongly scented, tubular, 1-3 mm long with a long, curved style when released. They are arranged in axillary clusters of 2-11 flowers. **Fruits** 2-3 cm long, warty, woody, ovate follicles, with a small beak. **Flowering** Autumn and winter. **Habitat** Sandy soils and swampy sites in heaths and open forests of southern Vic., southeastern SA and northeastern Tas. **Family** Proteaceae.

Hakea teretifolia   Dagger Hakea

Hardy, straggling shrub to 3 m high with rigid branches and whitish new growth. **Leaves** Needle-like, stiff and prickly, 15-70 mm long and 1-2 mm wide. **Flowers** White to yellowish, covered with white hairs, tubular, 4-6 mm long with deeply-cut, rolled-back petals and a long, curved, style when released. They are arranged in profuse, short, stalkless, axillary clusters of about 6 flowers. **Fruits** 2-3 cm long and 6-8 mm wide, woody, dagger-shaped follicles with a horny base. **Flowering** Spring and summer. **Habitat** Damp, sandy soils in coastal heaths and scrubs of the coast and ranges of southeastern Qld, NSW, Vic. and northeastern Tas. **Family** Proteaceae.

Hakea sericea   Bushy Needlewood. Silky Hakea

Stiff, erect, slender or bushy shrub to 3 m high with silky-hairy young branches. **Leaves** Stiff, needle-like, prickly, 15-80 mm long and 1-2 mm wide, at right angles to the stem. **Flowers** White or pink, scented, tubular, 4-5 mm long with deeply-cut, rolled-back, hairy petals and long, curved styles when released. They are arranged in axillary clusters of 1-7 flowers. **Fruits** 2-4 cm long and 20-25 mm wide, ovoid, wrinkled and nut-like follicles, with a double-pointed beak 3-4 mm long. **Flowering** Winter and spring. **Habitat** Heaths and open forests of the coast and adjacent ranges south of Coffs Harbour in NSW, Vic., southeastern SA and northeastern Tas. **Family** Proteaceae.

Hakea dactyloides   Finger Hakea

Erect, bushy shrub, to 3 m high with silky new growth. **Leaves** Stiff with conspicuous veins, linear to lanceolate or obovate, 5-10 cm long and 5-20 mm wide, flat and pointed. **Flowers** White or cream, tubular, 2-4 mm long with a long, curved style when released. They are arranged in crowded axillary clusters or short racemes. **Fruits** Hairless, warty, woody, ovoid follicles, 2-3 cm long and 15-18 mm wide with a small, straight beak. **Flowering** Spring and early summer. **Habitat** Sandy soils in open forests and heaths of the coast and tablelands of southeastern Qld, NSW and southeastern Vic. **Family** Proteaceae.

Hakea rostrata   Beaked Hakea

Stiff, spreading or rounded shrub to 5 m high with white-hairy branches. **Leaves** Needle-like, 2-15 cm long and 1-2 mm wide, curling upwards with long, sharp points. **Flowers** White, tubular, 3-6 mm long with curled-back petals and long, curved styles when released. They are arranged in axillary clusters of 1-10 flowers. **Fruits** Woody, wrinkled and rough, globular follicles, 2-5 cm long, with a long, curved-back beak. **Flowering** Spring. **Habitat** Heaths and open forests of southeastern SA, western Vic. and northeastern Tas. **Family** Proteaceae.

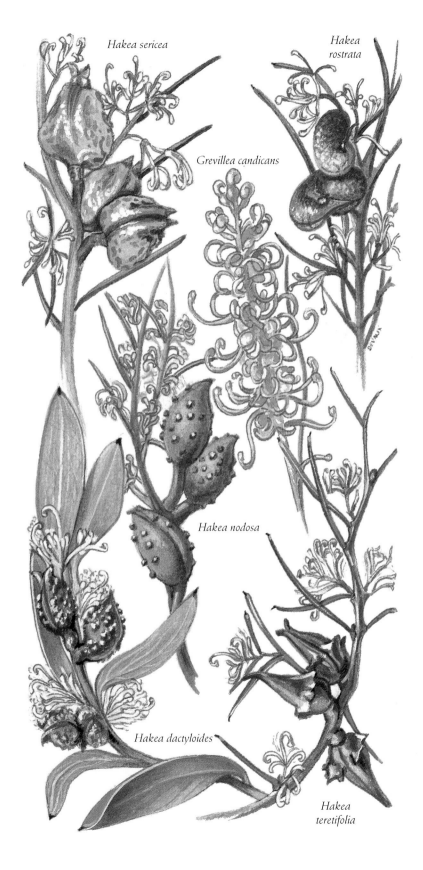

Hakea sericea

Hakea rostrata

Grevillea candicans

Hakea nodosa

Hakea dactyloides

Hakea teretifolia

Hakea bakeriana

Rounded shrub to 1.5 m high with slightly hairy young branches. **Leaves** Rigid, needle-like and pointed, 2-7 cm long and 1-2 mm wide. **Flowers** Pink to crimson, tubular, 8-15 mm long with rolled-back petals and curved styles, about 4 cm long when released. They are arranged in crowded axillary clusters of 6-12 flowers well down the branches. **Fruits** Rough and woody, ovoid follicles, 5-7 cm and 3-4 cm diameter with a short beak. **Flowering** From late autumn to winter. **Habitat** Sandy soils in coastal heaths and open forests of central eastern and southeastern NSW. **Family** Proteaceae.

Hakea purpurea   Crimson Hakea

Erect, rigid, rounded shrub to 3 m high with silky hairs on new growth. **Leaves** Needle-like, 3-10 cm long and 1-2 mm wide, usually divided at the ends into several short, pointed tips. **Flowers** Bright-red, tubular, 7-10 mm long, with rolled-back petals and long, curved, styles when released. They are arranged in axillary clusters of 10-12 flowers. **Fruits** Woody, rough, ovoid follicles, 3-4 cm long and 15-20 mm diameter, with a very small beak. **Flowering** Spring. **Habitat** Open and sparsely wooded country west of the Great Dividing Range in the southern half of Qld and northern NSW. **Family** Proteaceae.

Telopea truncata   Tasmanian Waratah

Erect, spreading shrub, usually to 3 m high, rarely a multi-stemmed tree to 8 m, with soft, rusty hairs on young branches. **Leaves** Alternate, 3-18 cm long and 5-22 mm wide, oblong to obovate with thick, slightly curved-back margins, dull dark-green above, covered with rusty hairs below. **Flowers** Bright-red, rarely yellow, tubular, to 25 mm long with 4 curled-back lobes and long, curved, protruding styles. They are arranged in dense, terminal, globular heads of 10-35 flowers, 3-8 cm across. **Fruits** Dark-brown woody follicles, 35-65 mm long, containing winged seeds. **Flowering** Spring and summer. **Habitat** Widespread in montane heaths and wet forests, 600-1200 m high, in Tas. **Family** Proteaceae.

Petrophile linearis   Narrow-leaf Cone Bush. Drumsticks

Erect, open shrub to 80 cm high. **Leaves** Alternate, narrow-obovate to sickle-shaped with sharp points, thick, grey-green, 5-12 cm long and 2-10 mm wide. **Flowers** Pink to violet and yellow with grey tips, covered with soft hair, narrow tubular, 20-35 mm long with 4 curled-back lobes when open. They are arranged in crowded, terminal or axillary, spidery, ovoid to globular heads. **Fruits** Ovoid cones, about 25 mm long, containing nuts about 3 mm long. **Flowering** Spring. **Habitat** Common on sand plains and open woodlands in southwestern WA. **Family** Proteaceae.

Lomatia myricoides   Long-leaf or River Lomatia

Erect shrub or small tree, 2-5 m high with rusty-hairy buds. **Leaves** Alternate, linear to narrow-oblong or lanceolate, stalkless, pointed, with entire or coarsely-toothed margins, paler below, 5-20 cm long and 5-20 mm wide. **Flowers** Cream to greenish-yellow, 8-10 mm long, with 4 separate, curled-back lobes and a long, protruding style when released. They are arranged in loose, spidery, axillary racemes, 5-10 cm long. **Fruits** Thin-walled woody follicles, 2-3 cm long. **Flowering** Summer. **Habitat** Widespread along watercourses and in sclerophyll forests on sandy soils to 1000 m along the coast and ranges of central and southern NSW and eastern Vic. **Family** Proteaceae.

Lomatia silaifolia   Wild Parsley. Crinkle Bush

Erect shrub to 2 m high. **Leaves** Alternate, pinnately divided 2 or 4 times into stalkless, linear or lanceolate, sharply-toothed segments. The whole leaf is 10-35 cm long with segments 3-20 mm wide. **Flowers** White to cream, 14-16 mm long with 4 separate, curled-back lobes and a long, curved, protruding style when released. They are arranged in loose, spidery, terminal racemes, 20-30 cm long. **Fruits** Thin-walled woody follicles, 3-5 cm long. **Flowering** Summer. **Habitat** Widespread in sandstone heaths, dry sclerophyll forests and woodlands of the coast and tablelands in southeastern Qld, northeastern and central eastern NSW. **Family** Proteaceae.

Hakea purpurea

Lomatia silaifolia

Hakea bakeriana

Petrophile linearis

Telopea truncata

Lomatia
myricoides

Dendrobium tetragonum — Tree Spider Orchid

Perennial herb growing on rocks or trees, with thick, 4-angled, succulent, swollen stems, 6-60 cm long and 7-9 mm diameter, spreading to pendant. **Leaves** Thin, stiff, dark-green, broad-lanceolate or elliptic, 3-10 cm long, 15-25 mm wide, with 2-5 leaves growing at the tip of each stem. **Flowers** Yellow-green with red-brown markings, spidery with 5 long, narrow lobes 2-5 cm long and a large, broad labellum 8-12 mm long and 6-8 mm wide, embracing the column of fused stamens and style. They are in terminal clusters of 1-5 flowers on long, thin stems. **Flowering** Spring. **Habitat** Rocks and trees in moist, shady areas in rainforests and along streams of the coast and adjacent ranges of Qld, and northern and central eastern NSW. **Family** Orchidaceae.

Diuris aurea — Golden Diuris

Robust, perennial herb to 60 cm high. **Leaves** Narrow-linear, ribbed and deeply channelled, arising from the base of the stem, 10-30 cm long and 4-6 mm wide. **Flowers** Golden yellow to orange with brown markings, 2 round lobes 15-22 mm across stand erect, 2 long narrow lobes 17-24 mm long project below the column of separate stamens and style, the labellum is deeply 3-lobed and 10-16 mm long. They are arranged in a terminal raceme of 2-5 flowers on stalks 30-60 cm long. **Flowering** Late winter and spring. **Habitat** Widespread in moist grassy sites in sclerophyll forests along the coast and adjacent ranges of Qld and northern and central eastern NSW. **Family** Orchidaceae.

Diuris longifolia — Common Donkey Orchid

Robust perennial herb to 45 cm high. **Leaves** Linear to lanceolate, channelled, arising from the base of the stem, 7-20 cm long and 4-12 mm wide. **Flowers** Yellow with brown and purple markings, 2 broad lobes about 16 mm long stand erect, 2 long narrow lobes 18-25 mm long project below the column of separate stamens and style, the labellum is deeply 3-lobed. They are arranged in a terminal raceme of 1-8 flowers on long stalks. **Flowering** Winter and spring. **Habitat** Widespread in sheltered sites along the coast of southern NSW, Vic., SA, southwestern WA and Tas. **Family** Orchidaceae.

Diuris punctata — Purple Donkey Orchid. Double Tails

Robust perennial herb to 60 cm high. **Leaves** Linear, channelled, arising from the base of the stem, 15-30 cm long and 2-5 mm wide. **Flowers** Mauve to yellow or white with purple infusions, sometimes dotted, 3 broad lobes 1-3 cm long stand erect, 2 very long narrow lobes 3-9 cm long project below the column of separate stamens and style, the labellum is deeply 3-lobed, the middle lobe being much broader and longer, 9-15 mm long and 10-15 mm wide. They are arranged in a terminal raceme of 1-10 flowers on stalks 30-60 cm long. **Flowering** Spring and early summer. **Habitat** Widespread in moist grasslands, sclerophyll forests and heaths in Qld, NSW, Vic. and southeastern SA. **Family** Orchidaceae.

Cryptostylis subulata — Slipper or Large-tongue Orchid

Upright perennial herb to 80 cm high. **Leaves** Narrow-ovate to lanceolate, stiff and erect, 5-20 cm long and 1-3 cm wide. **Flowers** Yellow-green with a large red or brown tubular labellum, 15-35 mm long and 5-10 mm wide, with long, narrow lobes, 15-30 mm long and about 3 mm wide spreading out behind the column. They are arranged in a tall raceme of 2-14 flowers on a stalk 15-80 cm long. **Flowering** Spring and summer. **Habitat** Moist and swampy sites, sclerophyll forests and woodlands of the coast and adjacent ranges of Qld, NSW, Vic., southeastern SA and Tas. **Family** Orchidaceae.

Caladenia arenaria (syn. C. patersonii) — Common Spider Orchid

Upright, hairy perennial herb to 40 cm high. **Leaf** (solitary) Oblong to linear or lanceolate, arising from the base of the plant, hairy, reddish at the base, 10-20 cm long and 5-18 mm wide. **Flowers** White to yellow or crimson with 5 very long and narrow spreading lobes, 4-10 cm long, and a curved labellum with toothed margins, about 18 mm long and 10 mm wide, with 4-6 rows of projecting calli in front of the column. They are arranged in small terminal racemes of 1-6 flowers, to 40 cm high. **Flowering** Spring. **Habitat** Widespread in forested areas and on sandhills in NSW, Vic., SA and Tas. **Family** Orchidaceae.

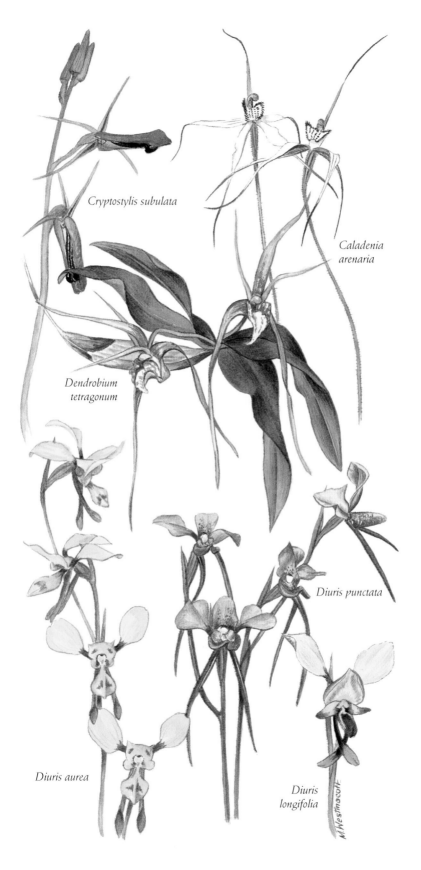

Cryptostylis subulata

Caladenia arenaria

Dendrobium tetragonum

Diuris punctata

Diuris aurea

Diuris longifolia

M. Westmacott

Caleana major — Large or Flying-duck Orchid

Upright, slender, perennial herb to 60 cm high. **Leaf** (solitary) Usually prostrate, narrow-lanceolate, grass-like, arising from and sheathing the base of the plant, 6-12 cm long and 8 mm wide, often spotted. **Flowers** Red-brown to purple, with green lower sepals, resembling a duck in flight, about 3 cm long with narrow lobes about 2 cm long and an upright, ovoid labellum curled into the shape of a head, the column of fused stamens and style resembling a duck's body. They are in terminal clusters of 1-5 flowers on a wiry stalk to 40 cm high. **Flowering** Spring and summer. **Habitat** Widespread in sclerophyll forests swampy and coastal heaths along the coast and ranges of Qld, NSW, Vic., southeastern SA and Tas. **Family** Orchidaceae.

Cryptostylis erecta — Bonnet or Tartan-tongue Orchid

Upright perennial herb to 80 cm high. **Leaves** Ovate to lanceolate, stiff and erect, arising from the base of the plant, dark-green above and reddish purple below, 5-13 cm long and 1-3 cm wide. **Flowers** Green with reddish-purple markings, with a large labellum 20-35 mm long and 5-15 mm wide forming an erect hood, with long narrow lobes behind. They are arranged in a tall raceme of 2-12 flowers on a stalk 15-80 cm long. **Flowering** Spring, summer and autumn. **Habitat** Widespread in heaths and open forests along the coast and adjacent ranges of southeastern Qld, NSW and southeastern Vic. **Family** Orchidaceae.

Pterostylis nutans — Nodding Greenhood

Upright perennial herb to 40 cm high. **Leaves** In a rosette of 3-6 around the base of the plant, oblong to elliptic or ovate, 15-90 mm long and 1-2 cm wide with wavy margins. **Flowers** Translucent green, sometimes tipped with brown, 2-3 cm long, strongly nodding, with a large hood over the column, a curved, pointed labellum and 2 narrow lobes project either side of the hood. They are solitary on a long stalk. **Flowering** Early winter and spring. **Habitat** Widespread in sandy soils in moist sites in dry sclerophyll forests and coastal scrub along the coast and adjacent ranges of Qld, NSW, Vic., southeastern SA and Tas. **Family** Orchidaceae.

Pterostylis grandiflora — Superb or Cobra Greenhood

Upright perennial herb to 35 cm high. **Leaves** In a rosette of 4-9 around the base of the plant, separated from the flowering stem, ovate to heart-shaped, 7-30 mm long and 6-15 mm wide, with scattered sheathing leaves along the stem, narrow-ovate to lanceolate, 15-60 mm long and 3-8 mm wide. **Flowers** Translucent white and green with red-brown markings, to 4 cm long with a large hood over the column, a curved labellum and 2 long narrow lobes projecting on either side of the hood. They are usually solitary on a stalk to 40 cm long, bearing 4-0 stem leaves. **Flowering** Autumn and winter. **Habitat** Widespread in moist, shaded gullies and forests in eastern Qld, eastern NSW, southern Vic. and northeastern Tas. **Family** Orchidaceae.

Orthoceras strictum — Horned or Bird's-mouth Orchid

Upright perennial herb to 60 cm high. **Leaves** Grass-like, channelled, arising from and sheathing the base of the plant, 5-30 cm long and 3 mm wide. **Flowers** Green to brown, hooded with 2 long projecting lobes and a curved labellum 8-12 mm long, arranged in a narrow, rigid raceme of 1-9 flowers. **Flowering** Late spring and summer. **Habitat** Wet, sandy soils in heaths and forests in southeastern Qld, eastern NSW, Vic., SA and Tas. **Family** Orchidaceae.

Eriochilus cucullatus — Parson's Bands

Upright slender perennial herb to 25 cm high with hairy stems. **Leaf** (solitary) Ovate, grey-green, arising from the base of the plant, 5-35 mm long and 8 m wide. **Flowers** Pink or white, perfumed, hairy, with 2 ovate lower spreading lobes, 10-17 mm long, 2 linear upright lobes and a curled labellum 5-8 mm long with bristly pink or bronze calli, arranged in a terminal raceme of 1-5 flowers, to 25 cm high. **Flowering** Summer and autumn. **Habitat** Widespread in open forests and heaths from sea-level to subalpine regions in all states except WA and the NT. **Family** Orchidaceae.

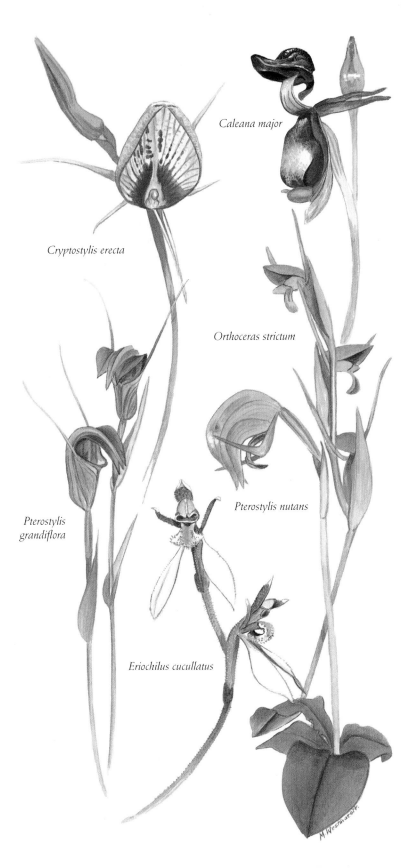

Caleana major

Cryptostylis erecta

Orthoceras strictum

Pterostylis nutans

Pterostylis grandiflora

Eriochilus cucullatus

M. Westmacott.

Caladenia gracilis (syn. C. angustata)   Musky Caladenia

Upright, sparsely hairy perennial herb to 45 cm high with a musky perfume. **Leaf** (solitary) Linear, grass-like, arising from the base of the plant, slightly hairy, 6-30 cm long and 6 m wide. **Flowers** Chocolate-purple and hairy outside, pink or cream inside, 20-35 mm across with large hood-like lobe arched over the column of fused stamens and style, and 4 long narrow lobes fanned out in front of the column. The labellum is ovate, 7-9 mm long and to 7 mm wide with short calli in 4 rows and toothed margins on the mid-lobe. They are solitary or in terminal racemes of 2-6 flowers, up to 45 cm high. **Flowering** Spring. **Habitat** Forested slopes with stringy-barks and peppermints along the coast and ranges of NSW, Vic., southeastern SA and northeastern Tas. **Family** Orchidaceae.

Calochilus robertsonii   Purplish Beard Orchid

Upright perennial herb to 45 cm high. **Leaf** (solitary) Arises from the base of the plant. It is linear to lanceolate with a conspicuous channel, to 30 cm long and 13 mm wide. **Flowers** Greenish-brown with red or purple stripes, a long, bearded lower lip about 2 cm long extends from the column of fused stamens and style with a striped, hood-like lobe arching above, 2 curled wing-like lobes on either side of the column with 2 spreading lobes below. They are arranged in a terminal raceme of 1-5 flowers, to 45 cm high. **Flowering** Spring. **Habitat** Widespread in heaths and open forests in southeastern Qld, the coast, ranges and western slopes of NSW, Vic., southeastern SA, southwestern WA and Tas. **Family** Orchidaceae.

Calochilus campestris   Copper Beard Orchid

Upright perennial herb to 60 cm high. **Leaf** (solitary) Linear to lanceolate, rigid, fleshy, triangular in section with a conspicuous channel, arising from the base of the plant, 11-30 cm long and to 1 cm wide. **Flowers** Green with red or purple stripes, a bearded lower lip about 2 cm long extends from the column of fused stamens and style with a hood-like lobe arching above, and 2 curled wing-like lobes on each side of the column. They are arranged in a terminal raceme of 5-15 flowers, to 60 cm high. **Flowering** Spring and early summer. **Habitat** Widespread in moist sclerophyll forests on slopes and ridges along the coast and inland in southeastern Qld, NSW, Vic., SA and Tas. **Family** Orchidaceae.

Corybas diemenicus (syn. C. dilatatus) Veined or Stately Helmet Orchid

Dwarf perennial herb to 4 cm high. **Leaf** (solitary) Orbicular to heart-shaped, usually flat on the ground, 8-25 mm long and 8-20 mm wide. **Flowers** Dark reddish-purple, 15-30 mm long with a large hood with toothed margins over a short, stout, fleshy column and a tubular labellum, 14-18 mm long, widely flared with toothed margins. They are solitary on short stalks. **Flowering** Winter and early spring. **Habitat** Cool mountain gullies and moist, shaded, site in sclerophyll forests in the southern tablelands of NSW, Vic., southeastern SA, southwestern WA and Tas. **Family** Orchidaceae.

Lyperanthus nigricans   Red Beaks or Undertaker Orchid

Upright, stout and fleshy perennial herb to 30 cm high, the flowers and leaves turn black when dried. **Leaf** (solitary) Flat on the ground, broad-ovate, fleshy, 2-12 cm long and 2-8 cm wide. **Flowers** White to pink with purple-red stripes, about 4 cm across with a beak-like hood arching over a fringed labellum to 2 cm long with 2 rows of calli, an incurved column and 4 narrow spreading lobes, arranged in a terminal raceme of 2-8 flowers, to 20 cm high. **Flowering** Spring. **Habitat** Sandy heaths, sclerophyll forests and woodlands in southeastern Qld, coastal NSW, Vic., SA, southwestern WA and Tas. **Family** Orchidaceae.

Lyperanthus suaveolens   Brown Beaks

Upright perennial herb to 45 cm high with a vanilla-like perfume. **Leaf** (solitary) Linear to lanceolate with curled-back margins, arising from the base of the plant, stiff, leathery, 12-26 cm long and 1-2 cm wide, paler below. **Flowers** Red-brown to yellow-green, to 3 cm diameter with a narrow, beak-like hood over a 3-lobed labellum about 1 cm long with several rows of small, pale calli, an incurved column and 2 narrow spreading lobes above and below, arranged in a terminal raceme of 2-8 flowers. **Flowering** Spring. **Habitat** Widespread in scrubs, heaths and open forests of the coast and tablelands in southeastern Qld, NSW, eastern Vic. and northeastern Tas. **Family** Orchidaceae.

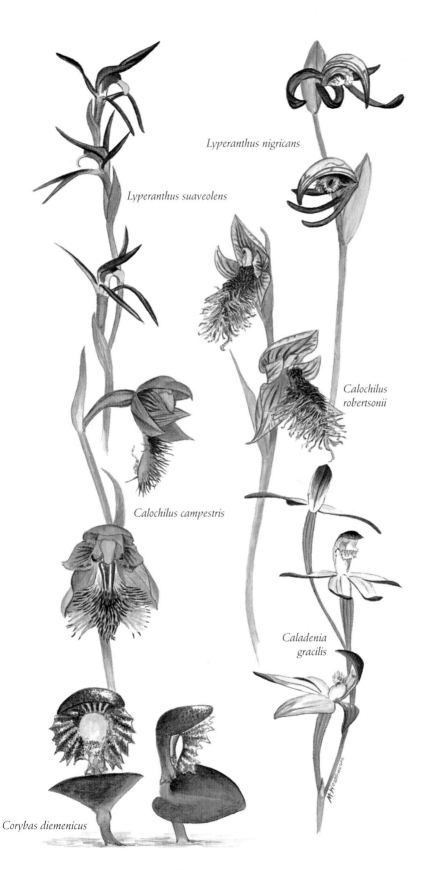

Lyperanthus nigricans

Lyperanthus suaveolens

Calochilus robertsonii

Calochilus campestris

Caladenia gracilis

Corybas diemenicus

Pterostylis longifolia
Tall Greenhood

Upright perennial herb to 40 cm high. **Leaves** In a rosette of 3-6 around the base of the plant, lanceolate, 2-4 cm long and 3-4 mm wide, usually separate from the flowering stem which has 5-8 alternate, broad-linear to lanceolate stem-clasping leaves, 1-10 cm long and 3-5 mm wide, absent when flowering. **Flowers** Green and white, 1-2 cm long with a large hood over the column and an oblong labellum, 4-6 mm long and 2-3 mm wide, that flicks back into the hood when touched, with 2 lobes below. They are arranged in a terminal raceme of 1-7 flowers, 15-40 cm high. **Flowering** Autumn and winter. **Habitat** Widespread in moist shady sites in sclerophyll forests and coastal scrubs along the coast and ranges of southeastern Qld, NSW, Vic., SA and Tas. **Family** Orchidaceae.

Prasophyllum flavum
Yellow Leek Orchid

Upright perennial herb to 90 cm high. **Leaf** (solitary) Sheaths the base of the stem for half its length, leaf and stem are to 60 cm long, dark purple to greenish black. **Flowers** Yellow-green with brown markings, sweetly perfumed, 15-20 mm across with a beaked hood above a short column and a concave labellum to 8 mm long with wrinkled margins, 2 concave spreading lobes on each side and one below. They are arranged in a long terminal raceme of 4-50 flowers. **Flowering** Spring and summer. **Habitat** Widespread in woodlands and forests of the coast and tablelands in high rainfall areas of southeastern Qld, NSW, Vic. and Tas. **Family** Orchidaceae.

Dendrobium schoeninum (syn. D. beckleri)
Pencil Orchid

Perennial herb with slender, erect or pendant stems 1-2 m long, growing on trees and rocks. **Leaves** Linear and fluted, fleshy, dark-green, erect, 2-16 cm long and 2-12 mm diameter. **Flowers** White or yellow-green with purple markings, about 25 mm across with 5 spreading narrow lobes on one side of the frilled, curved labellum, 2-3 cm long and 7-10 mm wide, which embraces the column of fused stamens and style. Arranged in small racemes of 1-4 flowers, 15-30 mm long. **Flowering** Spring. **Habitat** Rainforest margins and cliff faces, riverbanks and swamp margins of southeastern Qld and the northeastern NSW. **Family** Orchidaceae.

Dendrobium gracilicaule

Perennial herb with thick, erect to spreading, fleshy stems to 90 cm long, growing on trees and rocks. **Leaves** Lanceolate to ovate, leathery, dark-green, 5-13 cm long and 2-4 cm wide. **Flowers** Yellow with red-brown blotches outside, fragrant, about 15 mm across with 5 spreading lobes 10-11 mm long arching over the curved labellum which embraces the column of fused stamens and style. Arranged in slender racemes of 5-30 flowers, 5-12 cm long. **Flowering** Late winter and spring. **Habitat** Rainforests of coastal Qld and the north and central coast of NSW. **Family** Orchidaceae.

Caladenia deformis
Blue Fairies. Bluebeard Caladenia

Dwarf, slightly hairy perennial herb to 17 cm high. **Leaf** (solitary) Linear to lanceolate, arising from the base of the plant, thin, 4-10 cm long and 2-5 mm wide. **Flowers** Blue to violet, rarely white, pink or yellow, about 3 cm across with 5 spreading lobes and an erect, fringed labellum 10-15 mm long against a long, incurved column of fused stamens and style, usually terminal and solitary. **Flowering** Winter and spring. **Habitat** Damp, sheltered sites in heaths and mallee, of the central and southern coast and plains of NSW, Vic., SA, southwestern WA and northeastern Tas. **Family** Orchidaceae.

Chiloglottis pluricallata (syn. C. gunnii)
Common Bird Orchid

Low, tuberous, perennial herb to 8 cm high. **Leaves** arise from the base of the plant, there are 2, broad-ovate to oblong with short stalks, 5-10 cm long and 2-4 cm wide. **Flowers** Green to reddish-brown or purple, waxy, 2-4 cm long and 25-40 mm across resembling an open bird's mouth, the labellum has a few large black-tipped stalked glands in the centre. The flower is solitary on a thick stalk. **Flowering** Spring and early summer. **Habitat** Widespread in moist alpine and sub-alpine forests to about 1200 m along the coast and tablelands of NSW, Vic. and Tas. **Family** Orchidaceae.

Dendrobium schoeninum

Caladenia deformis

Dendrobium gracilicaule

Prasophyllum flavum

Pterostylis longifolia

Chiloglottis pluricallata

M.Westmacott

Dipodium punctatum   Hyacinth Orchid

Upright, perennial saprophytic herb to 90 cm high. **Leaves** Reduced, at the base of the flowering stem, ovate to broad-ovate, 7-30 mm long,. **Flowers** Pink to dark mauve, usually spotted, 1-3 cm across with 5 regular spreading lobes, 15-17 mm long, a cylindrical central column and protruding labellum. They are arranged in a terminal thick, fleshy raceme of 14-60 flowers, 40-100 cm long. **Flowering** Summer and autumn. **Habitat** Lives on dead organic matter, commonly in sandy sites in wet sclerophyll forests and woodlands of the coast, ranges and western slopes of Qld, NSW, Vic., southeastern SA and northeastern Tas. **Family** Orchidaceae.

Dendrobium kingianum   Pink Rock Orchid

Upright perennial herb to 30 cm high, with erect to spreading succulent stems, growing on rocks, rarely trees, or on the ground. **Leaves** Dark-green, narrow-elliptic to narrow-obovate, 2-13 cm long and 1-2 cm wide, at the end of the pseudo-bulb. **Flowers** Pink to mauve, rarely white, 12-30 mm across with 5 regular spreading, overlapping lobes and a blotchy labellum embracing the column, arranged in slender axillary racemes 7-20 cm long of 2-15 flowers. **Flowering** Winter and spring. **Habitat** Forms extensive clumps in open forests of the coast and ranges to about 1000 m altitude, from southeastern Qld to central eastern NSW. **Family** Orchidaceae.

Dendrobium bigibbum   Cooktown Orchid

Floral emblem of Queensland. A robust epiphytic perennial herb with pseudo-bulbs to 1 m long. **Leaves** Scattered along the upper part of the stem, dark-green and lanceolate, 8-15 cm long and 35 mm wide. **Flowers** Pink to mauve, rarely white, to 5 cm across with 5 regular spreading, overlapping lobes and a labellum embracing the column, arranged in slender racemes of 3-15 flowers, to 40 cm long. **Flowering** Most of the year. **Habitat** Grows on trees or rocks in forests of northeastern Qld. **Family** Orchidaceae.

Phaius tankervilliae

Upright perennial herb to 2 m high. **Leaves** Broad-lanceolate, folded around the base of the plant, 30-120 cm long and 25-65 mm wide. **Flowers** Brownish red inside and white outside, 7-10 cm across with 5 regular spreading lobes and a labellum about 5 cm long and 45 mm across with yellow and red markings, curled into a cylinder surrounding the column. They are arranged in 1-2 erect, terminal racemes of 8-14 flowers, 60-210 cm long. **Flowering** Spring. **Habitat** Grows among grasses and sedges mainly in wet coastal sites of Qld and far northeastern NSW. **Family** Orchidaceae.

Caladenia carnea   Pink Fingers

Upright, slender, perennial herb to 24 cm high. **Leaf** (solitary) Arises from the base of the plant, narrow-linear to lanceolate, 4-20 cm long and about 4 mm wide, sparsely hairy. **Flowers** White to pink, sweet or musk-scented, 2-4 cm across with 5 spreading lobes, one standing erect behind the column, and a 3-lobed, fringed labellum, to 1 cm long and 1 cm wide, with club-like calli in 2 rows. They are solitary or in small terminal racemes of 1-3 flowers, to 24 cm high. **Flowering** Spring and early summer. **Habitat** Sandy areas in heaths, sclerophyll forests and woodlands, along the coast and adjacent ranges of southeastern Qld, NSW, Vic. and SA. **Family** Orchidaceae.

Caladenia latifolia   Pink Fairy

Upright perennial herb to 40 cm high. **Leaf** (solitary) Arises from the base of the plant, very hairy, broad-lanceolate, 4-20 cm long and to 3 cm wide. **Flowers** White to pink, 25-35 mm across with 5 spreading lobes and a 3-lobed labellum, 6-7 mm long, with entire or toothed margins with 3-4 pairs of hairy clubbed calli, standing erect against the column. They are solitary or in small terminal racemes of 1-4 flowers, to 40 cm high. **Flowering** Spring and early summer. **Habitat** Light sandy soils, particularly amongst tea-trees in coastal Vic., SA, southwestern WA and Tas. **Family** Orchidaceae.

*Dipodium
punctatum*

*Caladenia
carnea*

*Dendrobium
kingianum*

Dendrobium bigibbum

*Caladenia
latifolia*

*Phaius
tankervilliae*

M. Westmacott

Caladenia catenata (syn. C. alba)   White Fingers. White Caladenia

Upright perennial herb to 30 cm high with wiry stems. **Leaf** (solitary) Arises from the base of the plant, sparsely hairy, linear to lanceolate, 8-20 cm long and 2-4 cm wide. **Flowers** White or pink with yellow markings, 2-5 cm across with 5 spreading lobes, one standing erect behind the column, and a toothed or fringed, 3-lobed labellum in front of the column. They are usually solitary or in small terminal racemes of 2-6 flowers, to 30 cm high. **Flowering** Winter and spring. **Habitat** Open forests and shrublands of the coastal and ranges of Qld, NSW, western Vic and Tas. **Family** Orchidaceae.

Dendrobium speciosum   Rock Lily. King Orchid

Robust perennial herb growing on rocks or rarely trees, with thick, succulent stems 8-40 cm long and 2-6 cm diameter, with 2-5 leaves at the tip. **Leaves** Broad oval to oblong, thick and leathery, dark-green, 4-25 cm long and 2-8 cm wide. **Flowers** White or yellow, strongly perfumed, to 25 mm across with 5 spreading lobes 20-40 mm long and a purple-striped labellum, 9-13 mm long and 9-13 mm wide, embracing the column, arranged in dense terminal racemes of 20-115 flowers, 20-70 cm long. **Flowering** Winter and spring. **Habitat** Grows on rocks in sclerophyll forests and sometimes on rainforest trees of the coast and tablelands in Qld, NSW and south-eastern Vic. **Family** Orchidaceae.

Dendrobium striolatum   Streaked Rock Orchid

Robust perennial herb with wiry, spreading to pendant stems to 50 cm long, growing on rocks. **Leaves** Cylindrical, narrow, pointed, succulent, shallowly furrowed, spreading or pendant, 4-14 cm long and 2-3 mm diameter. **Flowers** Creamy-yellow with red-brown stripes, 2-3 cm across with 5 spreading lobes and a white frilled labellum, 7-10 mm long and 3-5 mm wide embracing the column. They are solitary or rarely 2-3 on long stalks. **Flowering** Spring. **Habitat** Grows on exposed rocks in sclerophyll forests along the coast and tablelands, to about 1000 m in southeastern Qld, central and southern NSW, southeastern Vic. and northeastern Tas. **Family** Orchidaceae.

Sarcochilus falcatus   Orange Blossom Orchid

Semi-pendant perennial herb with short stems, 1-8 cm long, growing on trees or rarely on rocks. **Leaves** Stiff, oblong to lanceolate, slightly channelled, arising from the base of the plant, 2-16 cm long and 8-22 mm wide. **Flowers** White to cream with orange streaks on the labellum, sweetly perfumed, 2-5 cm across with 5 spreading lobes and a 3-lobed labellum, 3-6 mm long, embracing the column. They are arranged in pendulous axillary racemes of 1-15 flowers, 1-13 cm long. **Flowering** Winter and spring. **Habitat** Widespread, growing on rainforest trees, particularly *Acacia melanoxylon*, and sometimes on rocks, from about 300 m to 1200 m along the coastal ranges and tablelands of Qld, NSW and southeastern Vic. **Family** Orchidaceae.

Sarcochilus fitzgeraldii   Ravine Orchid

Semi-pendant perennial herb with branching stems 8-100 cm long, growing on rocks and rarely trees, often forming large mats. **Leaves** Flaccid, channelled, broad-lanceolate to linear, stem-clasping, 6-20 cm long and 10-15 mm wide. **Flowers** White to pink with crimson blotches in the centre, sweetly perfumed, 20-35 mm across with 5 spreading lobes and a 3-lobed labellum, 5-6 mm long, embracing the column. They are arranged in sturdy axillary racemes of 4-15 flowers, 10-20 cm long. **Flowering** Spring. **Habitat** Grows on rocks and sometimes on the bases of trees in moist, shady ravines in subtropical rainforests from about 500-700 m altitude, along the eastern slopes of the Great Dividing Range in southern Qld and northern NSW. **Family** Orchidaceae.

Glossodia major   Wax-lip Orchid

Upright, slender, hairy, perennial herb to 32 cm high. **Leaf** (solitary) Oblong to lanceolate, prostrate or nearly so, arising from the base of the plant, hairy, 3-15 cm long and 1-2 cm wide, with a vanilla perfume when dry. **Flowers** Purple to mauve, rarely white, 5-6 cm across with 5 spreading lobes and a white-based labellum, 8-11 mm long, against the column, solitary or in pairs on long stalks. **Flowering** Winter and spring. **Habitat** Widespread on sandy soils in sclerophyll forests, woodlands and coastal heaths along the coast, ranges and inland slopes in Qld, NSW, Vic., southeastern SA and Tas. **Family** Orchidaceae.

Sarcochilus falcatus

*Dendrobium
striolatum*

*Sarcochilus
fitzgeraldii*

*Caladenia
catenata*

Glossodia major

Dendrobium speciosum

Caladenia gemmata   Blue China Orchid

Upright hairy perennial herb to 25 cm high. **Leaf** (solitary) Arises from the base of the plant, hairy, oval, dark brown below, 30-35 mm long. **Flowers** Dark blue, rarely white, 3-7 cm across with 5 spreading lobes and an oval labellum studded with club-like calli in front of the column. They are solitary or in small terminal racemes of 2-6 flowers. **Flowering** Winter and spring. **Habitat** Widespread in southwestern WA. **Family** Orchidaceae.

Thelymitra aristata   Scented or Great Sun Orchid

Upright slender perennial herb to 1 m high. **Leaf** (solitary) Narrow-linear to lanceolate, stem-clasping, thick, fleshy and channelled, 10-35 mm long and 5-22 mm wide, often blotched. **Flowers** Blue to pink-mauve, perfumed, 2-6 cm across with 5 spreading lobes and similar spreading labellum, the column is hooded with a yellow lobe and 2 hairy lateral lobes. They are arranged in a terminal raceme of 5-30 flowers to 80 cm high. **Flowering** Winter and spring. **Habitat** Widespread in heaths and open forests of the eastern Qld, the coast and tablelands of NSW, southern Vic., southeastern SA and Tas. **Family** Orchidaceae.

Thelymitra ixioides   Spotted Sun Orchid

Upright, slender, perennial herb to 60 cm high. **Leaf** (solitary) Narrow-linear to lanceolate, stem-clasping, ribbed, to 20 cm long and 1 cm diameter. **Flowers** Pale blue to mauve, pink or white, spotted with deep blue, 3-6 cm across with 5 spreading lobes and similar spreading labellum, the column is hooded with a yellow lobe and 2 hairy lateral lobes. They are arranged in a terminal raceme of 3-9 flowers on slender stalks to 60 cm high. **Flowering** Winter, spring and summer. **Habitat** Widespread in temperate heaths and open forests from the coast to the subalps in eastern Qld, NSW, Vic., southeastern SA and Tas. **Family** Orchidaceae.

Thelymitra media   Tall Sun Orchid

Upright perennial herb to 90 cm high. **Leaf** (solitary) Narrow-linear to lanceolate, stem-clasping, channelled, often ribbed and fleshy, 15-30 cm long and up to 18 mm wide. **Flowers** Blue or purple, with deep blue markings, 15-35 mm across with 5 spreading lobes and similar spreading labellum, the column is erect but not hooded, with a yellow mid-lobe with several rows of upturned calli, and 2 lateral lobes. They are arranged in a terminal raceme of 5-25 flowers on slender stalks to 90 cm high. **Flowering** Spring and early summer, particularly after fire. **Habitat** Low-lying areas in forests and scrub of the coast and adjacent ranges in central and southern NSW, Vic. and northeastern Tas. **Family** Orchidaceae.

Thelymitra venosa   Large-Veined Sun Orchid

Upright perennial herb to 75 cm high. **Leaf** (solitary) Broad linear, stem-clasping, channelled, thick and fleshy, to 30 cm long and 7 mm wide. **Flowers** Blue with dark blue veins, occasionally pink or white, 20-25 mm across with 5 spreading lobes and similar spreading labellum, the column is winged. They are arranged in a terminal raceme of 1-6 flowers on slender stalks to 70 cm high, opening on dull days and remaining open at night. **Flowering** Summer. **Habitat** Widespread in swampy sites on exposed sandstone ledges, often at higher altitudes among cushion mosses and low vegetation along the coast and adjacent ranges in southeastern Qld, NSW, Vic., southeastern SA and Tas. **Family** Orchidaceae.

Ottelia ovalifolia   Swamp Lily

Aquatic, perennial or annual herb, rooted in the mud with submerged and floating leaves and flowers. **Leaves** Ovate to elliptical, 2-16 cm long and 3-6 cm wide, with stalks 60-120 cm long, arising from the submerged base of the plant. **Flowers** Cream to white or pale-yellow with a dark-red centre and yellow stamens, open, 3-6 cm across, with 3 overlapping rounded petals, solitary on long stalks within a 2-lobed green to purple stem leaf. **Flowering** Warmer months. **Habitat** Widespread in still and slowly flowing, shallow fresh water to about 1 m deep, usually with high levels of nutrient in all mainland states. **Family** Hydrocharitaceae.

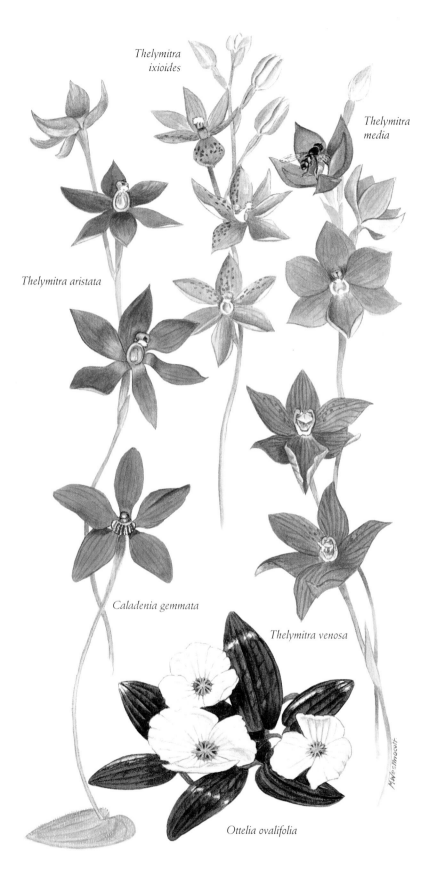

Thelymitra
ixioides

Thelymitra
media

Thelymitra aristata

Caladenia gemmata

Thelymitra venosa

Ottelia ovalifolia

Patersonia fragilis
Short Purple Flag. Fragile Iris

Erect, tufted, perennial herb to 50 cm high. **Leaves** Long, rigid, narrow-linear, deeply grooved, arising from the base of the stem, 15-80 cm long and 1-5 mm wide. **Flowers** Pale-purple to blue-violet, 20-35 mm across with 3 broad, spreading lobes and 3 yellow-tipped stamens. They are arranged in small terminal clusters on stalks 4-25 cm long, above 2 large brown to green sheathing bracts. **Fruits** Cylindrical capsules, 25-30 mm long with black, glossy seeds about 2 mm long. **Flowering** Spring. **Habitat** Widespread in damp sandstone and granite soils of southeastern Qld, eastern NSW, Vic., southeastern SA and northeastern Tas. **Family** Iridaceae.

Patersonia glabrata
Leafy Purple Flag. Wild Iris

Erect perennial herb to 80 cm high. **Leaves** Long, narrow-linear, arising from the base of the stem, 5-60 cm long and 2-5 mm wide. **Flowers** Pale purple, 25-50 mm across with 3 broad, spreading, delicate lobes and 3 yellow-tipped stamens, solitary or in terminal clusters of 2-3, on stems 10-30 cm long, above 2 brown, silky, sheathing bracts, 40-65 mm long. **Fruits** Cylindrical, 3-angled capsules, 2-4 cm long, with glossy brown seeds about 4 mm long. **Flowering** Spring. **Habitat** Widespread in sandy soils in woodlands and dry sclerophyll forests of the coast and ranges of Qld, NSW and southeastern Vic. **Family** Iridaceae.

Patersonia sericea
Silky Purple Flag. Bush Iris

Erect, perennial, densely-tufted herb with a silky-hairy flower stem, to 60 cm high. **Leaves** Long, linear, sword-shaped, tough, arising from the base of the stem, woolly at the base, 15-60 cm long and 2-6 mm wide. **Flowers** Purple, 3-4 cm across and 15-30 mm long, with 3 broad, spreading, extremely delicate and short-lived lobes and 3 yellow-tipped stamens. They are arranged in small terminal clusters on stems 3-55 cm long, above 2 brown, silky to smooth sheathing bracts, 2-6 cm long. **Fruits** Cylindrical to ovoid, 3-angled capsules, 15-30 mm long with brown seeds about 3 mm long. **Flowering** From winter to early summer. **Habitat** Widespread in sandy soils in dry sclerophyll forests, woodlands and heaths along the coast and ranges of Qld, NSW and southeastern Vic. **Family** Iridaceae.

Spyridium vexilliferum
Winged Spyridium

Wiry shrub to 1 m high with rusty hairs on young shoots. **Leaves** Alternate, widely separated, narrow-linear to obovate with curved-back margins, shiny-green with a depressed midvein, white or rusty hairy below, 8-15 mm long and 1-7 mm wide. **Flowers** In compact yellow button-like heads, 3-6 mm wide, of tiny 5-petalled individual **Flowers** surrounded by 1-4 white, velvety, petal-like floral leaves per head. **Fruits** Dark brown ovoid capsules about 25 mm long. **Flowering** Most of the year. **Habitat** Sandy heaths and open forests of central and western Vic., SA and Tas. **Family** Rhamnaceae.

Thysanotus multiflorus
Many-flowered Fringed Lily

Upright, tufted, perennial herb to 45 cm high. **Leaves** Narrow-linear, arising from the base of the plant, erect, rigid, 7-57 cm long and 2-5 mm wide. **Flowers** Blue to mauve, open, about 25 mm across with 3 broad, fringed, inner lobes and 3 narrower outer lobes, with 3 stamens. They are arranged in terminal clusters of 4-60 flowers, 7-70 cm long. **Fruits** Globular capsules with cylindrical seeds 2 mm long. **Flowering** Spring. **Habitat** Sandy soils, often in light woodlands and banksia scrubs in southwestern WA. **Family** Liliaceae.

Commelina cyanea
Prostrate perennial herb with ascending stems to 60 cm high. **Leaves** Alternate, ovate to lanceolate, 2-8 cm long and 5-15 mm wide. **Flowers** Bright blue, open, about 2 cm across with 3 separate lobes and 3 or 6 long stamens, on stalks with 1-3 flowers. **Fruits** Dry, 5-seeded capsules. **Flowering** From spring to autumn. **Habitat** Moist sites in forests and woodlands of the coast, tablelands and inland slopes of Qld, NSW and northern NT. **Family** Commelinaceae.

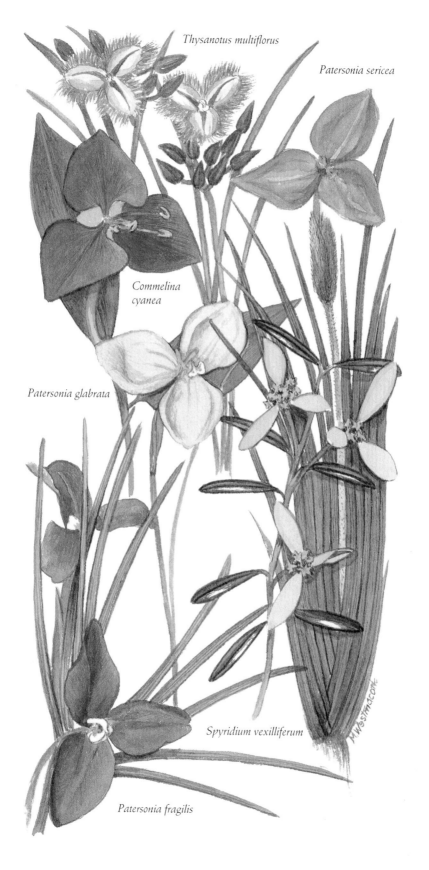

Thysanotus multiflorus

Patersonia sericea

Commelina cyanea

Patersonia glabrata

Spyridium vexilliferum

Patersonia fragilis

Cleome viscosa
Tickweed

Erect, slender annual herb with sticky stems to 1 m high. **Leaves** Alternate, divided into 3-5 lanceolate to narrow-elliptic leaflets, 1-7 cm long. **Flowers** Yellow with 4 spathulate petals, 7-16 mm long and 3-5 mm wide, grouped at one side of the stem, and 8-16 free stamens, 3-10 mm long, solitary and terminal on long stalks. **Fruits** Narrow-oblong pods, 3-10 cm long. **Flowering** Most of the year. **Habitat** Usually along watercourses in all mainland states except Vic. **Family** Capparaceae.

Epilobium billardierianum
Robust or Smooth Willow Herb

Upright perennial herb to 1 m high with hairy stems. **Leaves** Variable, linear to lanceolate or ovate, sometimes with coarsely-toothed margins, 5-80 mm long and 1-20 mm wide. **Flowers** Purplish pink or white, hairy, open, 10-36 mm across with 4 notched petals, 8 yellow stamens and a long protruding style. They are solitary or in clusters on long stalks in the upper leaf axils. **Fruits** Cylindrical, hairy capsules, 30-75 mm long with seeds about 1 mm long. **Flowering** Spring and summer. **Habitat** Widespread in moist areas of the coast and tablelands of southeastern Qld, NSW, Vic., SA, southwestern WA and Tas. **Family** Onagraceae.

Stylidium graminifolium
Grass Trigger Plant

Upright perennial herb with long flowering stems to 70 cm high. **Leaves** Linear to narrow-lanceolate with curved-back margins, sometimes finely toothed, tufted around the base of the plant, 5-30 cm long and 1-6 mm wide. **Flowers** Pink to magenta with a white centre, open, about 2 cm across, shortly tubular with 4 spreading petals and 2 stamens united into a long bent column that springs across when touched near the base to dust the insect with pollen, arranged in a loose raceme along a slender, hairy stem, 30-70 cm high. **Fruits** 2-valved ovoid capsules, 5-12 mm long. **Flowering** Spring and summer. **Habitat** Widespread in dry sclerophyll forests of the coast, tablelands and inland slopes in Qld, NSW, Vic., southeastern SA and Tas. **Family** Stylidiaceae.

Stylidium laricifolium
Tree Trigger Plant

Upright perennial herb or dwarf shrub with long flowering stems to 150 cm high. **Leaves** Narrow-linear, crowded, spirally arranged, 1-4 cm long and about 1 mm wide. **Flowers** Pink with a white centre, open, shortly tubular, 1-2 cm across with 4 spreading petals and 2 stamens united into a long, bent column that springs across when touched near the base to dust the insect with pollen. They are arranged in a loose raceme. **Fruits** 2-valved oblong capsules 8-12 mm long. **Flowering** Spring and early summer. **Habitat** Widespread in sclerophyll forests in mountainous and rocky sites of the coast, tablelands and inland slopes in southeastern Qld, NSW and southeastern Vic. **Family** Stylidiaceae.

Eschscholtzia californica
California Poppy

Upright annual herb to 50 cm high. **Leaves** Finely divided into numerous small flat linear lobes, 2-4 mm wide, on stalks to 12 cm long. **Flowers** Yellow to orange, open, 5-6 cm across with 4 overlapping rounded petals and numerous yellow stamens. They are solitary on long stalks. **Fruits** Cylindrical capsules 5-9 cm long and 4-5 mm across, exploding when ripe. **Flowering** Winter, spring and summer. **Habitat** Introduced from California, a widespread garden escape, naturalised in Qld, NSW, SA, WA and Tas. **Family** Papaveraceae.

Mitrasacme polymorpha

Upright perennial herb to 40 cm high. **Leaves** Opposite, oblong to lanceolate or elliptical with curved-back margins, hairy, 4-15 mm long and 1-6 mm wide. **Flowers** White, often with a yellow centre, open, 6-7 mm across, shortly tubular with 4 spreading lobes and 4 stamens, arranged in terminal clusters of 3-6 flowers, about 10 cm long. **Fruits** Small globular capsules, 2-3 mm diameter. **Flowering** Spring and summer. **Habitat** Widespread, often on sandy soils long the coast and tablelands of Qld, NSW and southeastern Vic. **Family** Loganiaceae.

*Stylidium
laricifolium*

Cleome viscosa

*Epilobium
billardierianum*

Mitrasacme polymorpha

Stylidium graminifolium

Eschscholtzia californica

M. Westmacott

Boronia falcifolia — Wallum Boronia

Low shrub to 1 m high with slender, angled, red-brown stems. **Leaves** Fleshy, channelled above, opposite, 5-25 mm long and 1-2 mm wide, often divided into 3 linear to cylindrical leaflets. **Flowers** Deep pink, open, 12-20 mm across with 4 pointed petals and 8 stamens, solitary or in threes in axillary clusters at the ends of the branches. **Fruits** Hairless capsules, exploding when ripe. **Flowering** Mainly in winter and spring. **Habitat** Damp coastal heaths in southeastern Qld and northern NSW. **Family** Rutaceae.

Boronia ledifolia — Showy, Sydney or Ledum Boronia

Bushy shrub to 1.5 m high with hairy young branches. **Leaves** Opposite, sometimes divided into 3-11 leaflets, oblong linear to narrow-elliptical, often with curved-back margins, shiny dark-green above, covered in white hairs below, 4-40 mm long and 1-7 mm wide, with a strong odour when crushed. **Flowers** Pink to red, 1-2 cm across, open with 4 pointed petals and 8 stamens, usually solitary on axillary stalks 6-12 mm long. **Fruits** Hairless capsules, exploding when ripe. **Flowering** Winter and spring. **Habitat** Coastal sandstone heaths and dry sclerophyll forests in southeastern Qld, NSW and southeastern Vic. **Family** Rutaceae.

Boronia microphylla — Small-leaved Boronia

Low compact shrub to 1 m high with sticky young branches. **Leaves** Opposite, thick, aromatic, pinnately divided into 5-15 linear to oblong or spathulate leaflets, 3-11 mm long and 1-4 mm wide. **Flowers** Deep pink, 10-15 mm across, open with 4 pointed petals and 8 stamens, in loose axillary clusters of 1-5. Flowers on stalks 3-10 mm long. **Fruits** Hairless capsules, exploding when ripe. **Flowering** Spring and summer. **Habitat** Sandy or granite soils in dry sclerophyll forests of the coast and tablelands in NSW and southeastern Qld. **Family** Rutaceae.

Boronia mollis — Soft Boronia

Bushy shrub to 2 m high with soft, downy branches. **Leaves** Opposite, pinnately divided into 3-9 oblong-elliptical to lanceolate leaflets with curled-back margins, downy below, the terminal leaflet is 15-40 mm long and 5-12 mm wide, the others much shorter and broader. **Flowers** Pale to deep pink, open, 15-20 mm across, with 4 pointed petals and 8 stamens, arranged in terminal or axillary clusters of 2-6 flowers on stalks 6-20 mm long. **Fruits** Smooth capsules, exploding when ripe. **Flowering** Winter and spring. **Habitat** Sandstone gullies in dry sclerophyll forests of the north and central coasts of NSW. **Family** Rutaceae.

Boronia pilosa — Hairy Boronia

Variable upright or prostrate aromatic shrub to 2 m high, usually covered with small hairs. **Leaves** Opposite, to about 15 mm long, pinnately divided into 3-7 pointed linear leaflets, 5-10 mm long. **Flowers** White to deep pink, open, to 15 mm across, waxy, with 4 pointed petals and 8 stamens, arranged in terminal or axillary clusters of 3-6 flowers. **Fruits** Capsules, exploding when ripe. **Flowering** Spring and summer. **Habitat** Damp coastal heaths in Vic., southeastern SA and Tas. **Family** Rutaceae.

Boronia pinnata — Pinnate Boronia

Erect shrub to 1.5 m high with strongly aromatic leaves and flowers. **Leaves** Opposite, pinnately divided into 5-11 thick, pointed, narrow-elliptic or narrow-oblong leaflets, 5-25 mm long and 1-3 mm wide with a strong, camphor-like odour when crushed. **Flowers** Bright to purplish pink, open, 1-2 cm across with 4 pointed petals and 8 stamens, arranged in loose axillary clusters of 3-8 flowers on slender stalks 6-30 mm long. **Fruits** Hairless capsules, exploding when ripe. **Flowering** Winter and spring. **Habitat** Coastal heaths and dry sclerophyll forests on sandstone in eastern NSW. **Family** Rutaceae.

Boronia microphylla

Boronia
falcifolia

Boronia pilosa

Boronia ledifolia

Boronia mollis

Boronia pinnata

M.Westmacott

Boronia algida
Alpine Boronia

Low, spreading shrub to 1.5 m high with red young branchlets. **Leaves** Opposite, thick, about 1 cm long, pinnately divided into 5-9 broad spathulate leaflets with curved-back margins, 3-7 mm long and 2-4 mm wide, rigid and shiny green. **Flowers** White to bright pink, open, 6-12 mm across, with 4 pointed waxy petals and 8 stamens, usually solitary in the terminal leaf axils. **Fruits** Hairless capsules, exploding when ripe. **Flowering** Spring and summer. **Habitat** Sandy soils in rocky and gravelly sites mostly at higher altitudes in NSW and southeastern Vic. **Family** Rutaceae.

Boronia coerulescens
Blue Boronia

Low spindly shrub to 60 cm high. **Leaves** Opposite, linear, blunt tipped, thick, 2-10 mm long and 0.5-1.5 mm wide. **Flowers** White to lilac or blue, open, 8-15 mm across with 4 waxy petals and 8 stamens, usually solitary in the upper leaf axils on stalks 2-4 mm long. **Fruits** Hairless capsules, exploding when ripe. **Flowering** Spring. **Habitat** Mallee communities and heaths, on sandy soils, in southeastern NSW, southwestern Vic., SA and southwestern WA. **Family** Rutaceae.

Boronia glabra
Blotched Boronia

Erect shrub to 1.5 m high with glandular, hairy branches. **Leaves** Opposite, blunt tipped, narrow-elliptical to oblong or linear with curved-back margins, 5-30 mm long and 1-6 mm wide. **Flowers** White to deep-pink with mauve blotches, open, 10-15 mm across, with 4 pointed petals and 8 stamens, solitary on axillary stalks 1-6 mm long. **Fruits** Hairy capsules, exploding when ripe. **Flowering** Winter and spring. **Habitat** Sandy and stony sites in open forests of the western slopes and plains in southeastern Qld and central and northern NSW. **Family** Rutaceae.

Boronia rosmarinifolia
Forest Boronia

Erect slender shrub to 1 m high with rusty brown hairy branchlets. **Leaves** Opposite in pairs, linear to narrow-oblong with curved-back margins, shiny dark-green above, whitish below, 6-40 mm long and 1-3 mm wide. **Flowers** Pale to deep pink, open, 10-18 mm across with 4 pointed petals and 8 stamens, solitary in the leaf axils on stalks 2-7 mm long. **Fruits** Hairless capsules, exploding when ripe. **Flowering** Winter and spring. **Habitat** Wet sites in heaths and open forests of southeastern Qld and northeastern NSW. **Family** Rutaceae.

Boronia whitei

Erect shrub to 1.5 m high with brown downy branches. **Leaves** Opposite, 10-35 mm long, pinnately divided into 3-9 narrow-linear leaflets 7-15 mm long and 1-2 mm wide, shiny-green above with curved-back margins and a depressed midvein, silky-hairy below. **Flowers** Pink-mauve to red, open, 15-20 mm across, with 4 pointed petals and 8 stamens, arranged in small clusters or solitary in the leaf axils. **Fruits** Hairless capsules, exploding when ripe. **Flowering** Winter and spring. **Habitat** Sandy, wet sites and granite soils in the tablelands of central and northern NSW. **Family** Rutaceae.

Zieria aspalathoides
Whorled Zieria

Erect shrub to 1 m high and 1.3 m across, with hairy branches. **Leaves** Whorled in groups of 3, linear to lanceolate with curled-under margins, 5-10 mm long and 1-3 mm wide. **Flowers** Pale to deep pink, open, 6-11 mm across with 4 pointed petals and 4 stamens, arranged in small axillary clusters of usually 3 flowers. **Fruits** Usually hairless, lobed capsules. **Flowering** From late winter to early summer. **Habitat** Rocky and sandy sites in heaths and dry sclerophyll forests of the tablelands, western slopes and plains in Qld, NSW and Vic. **Family** Rutaceae.

Boronia
coerulescens

Boronia
whitei

Boronia algida

Zieria aspalathoides

Boronia glabra

Boronia rosmarinifolia

M. Westmacott

Zieria arborescens
Stinkwood

Erect shrub or small tree to 10 m high with minutely hairy branches and a strong odour when handled.
Leaves Opposite, divided into 3 elliptic to narrow-elliptic or lanceolate leaflets often with slightly curved-back margins, smooth above, sometimes covered with small hairs below, 5-10 cm long and 6-30 mm wide.
Flowers White, downy, open, 6-14 mm across with 4 pointed petals and 4 stamens, solitary or in axillary clusters on fairly long stalks. **Fruits** Hairless, capsules with 4 seeds. **Flowering** Late winter and spring.
Habitat Moist forests and rainforest margins in southeastern Qld, eastern NSW, Vic. and Tas. **Family** Rutaceae.

Zieria laevigata

Erect, compact shrub to 1.5 m high with ridged branches. **Leaves** Opposite, divided into 3 linear-elliptic leaflets with curled-under margins, paler below with a prominent midvein, 15-50 mm long and 1-3 mm wide.
Flowers White to pale-pink, open, 6-10 mm across with 4 pointed petals and 4 stamens, solitary or in small axillary clusters. **Fruits** Hairless, lobed capsules. **Flowering** From late winter to spring. **Habitat** Sandy heaths and dry sclerophyll forests of the coast and tablelands in southeastern Qld and NSW. **Family** Rutaceae.

Zieria smithii
Sandfly Zieria

Erect shrub to 2 m high with rough, glandular branchlets and a strong odour when handled. **Leaves** Opposite, divided into 3 oblong to narrow-elliptic or lanceolate leaflets, with flat or slightly curved-back margins, 2-5 cm long and 4-10 mm wide. **Flowers** White to pale-pink, downy, open, 4-9 mm across with 4 pointed petals and 4 stamens, solitary or in loose axillary clusters of 6-70 flowers on stalks 1-2 cm long. **Fruits** Warty, hairless, lobed capsules. **Flowering** Spring and autumn. **Habitat** Moist forests, rainforest margins and cleared areas of the coast and tablelands in Qld, NSW, Vic. and Tas. **Family** Rutaceae.

Clematis aristata
Traveller's Joy. Austral Clematis

A vigorous woody climber to 6 m high. **Leaves** Opposite, often divided into 3, ovate to narrow-lanceolate or heart-shaped leaflets with irregularly toothed margins, 2-10 cm long and 10-45 mm wide on long twisted or coiled stalks. **Flowers** White, open, 3-5 cm across, star-like with 4 narrow, pointed lobes and many stamens, arranged in panicles on long stalks in the upper leaf axils. **Fruits** Silver-plumed ovate achenes, 2-4 cm long.
Flowering Spring and summer. **Habitat** Widespread in forests and moist, sheltered gullies of the coast and tablelands in southeastern Qld, NSW, Vic. and northeastern Tas. **Family** Ranunculaceae.

Clematis glycinoides
Headache Vine. Forest Clematis

A woody climber to 2.5 m high. **Leaves** Opposite, shiny dark-green, often divided into 3 ovate to oblong-lanceolate or heart-shaped leaflets, usually with 1-2 teeth on either side, 15-120 mm long and 1-8 cm wide on long twisted or coiled stalks. **Flowers** White or greenish, open, 3-4 cm across, star-like with 4 narrow, pointed lobes and numerous stamens, arranged in large clusters arising from the upper leaf axils on long stalks. **Fruits** Narrow-ovate silver-plumed achenes 3-4 cm long. **Flowering** Winter and spring. **Habitat** Widespread in a range of habitats, especially sclerophyll forests and rainforests of the coast and tablelands in Qld, NSW and southeastern Vic. **Family** Ranunculaceae.

Clematis microphylla
Small-leaved Clematis

A woody climber to about 3 m high. **Leaves** Opposite, divided 2 or 3 times into 3 narrow-linear to oblong leaflets, 2-50 mm long and 1-15 mm wide on long twisted or coiled stalks. **Flowers** White or creamy yellow, open, 2-5 cm across, star-like with 4 narrow lobes and many stamens, arranged in short panicles arising from the upper leaf axils on long stalks. **Fruits** Silver-plumed achenes, 2-4 cm long. **Flowering** Winter and spring.
Habitat Widespread in dry woodlands in highlands and inland plains in Qld, NSW, Vic., SA and northeastern Tas. **Family** Ranunculaceae.

Clematis aristata

Clematis microphylla

Clematis glycinoides

Zieria smithii

Zieria laevigata

Zieria arborescens

Tetratheca ciliata
Pink Eye

Small, slender, heath-like shrub to 1 m high with hairy, ridged branches. **Leaves** Alternate, opposite or in whorls of 3-5, ovate to orbicular, fringed with hairs, paler below, 2-20 mm long and 1-15 mm wide. **Flowers** Lilac to red or pink, rarely white, with a dark brown centre, open, 10-25 mm across with 4 spreading petals and 8-10 stamens, solitary or 2-3 together in the upper leaf axils. **Fruits** Obovate flattened capsules, 4-10 mm long. **Flowering** Spring. **Habitat** Sandy and loamy soils in heaths and sclerophyll forests of the southern coast, tablelands and slopes of NSW, Vic., southeastern SA and Tas. **Family** Tremandraceae.

Tetratheca ericifolia
Heath Pink Eye

Low, slender shrub to 60 cm high with bristly hairs on the branches. **Leaves** Alternate or in whorls of 4-6, narrow-linear with curled-under margins, rough, 3-10 mm long and 1-2 mm wide. **Flowers** Lilac to red or pink, rarely white, with a dark brown centre, open, 10-25 mm across with 4 spreading petals and 8 stamens, solitary or in pairs in the upper leaf axils. **Fruits** Obovate, beaked, flattened capsules 6-9 mm long and 2-4 mm wide. **Flowering** Mainly in spring and summer. **Habitat** Heaths, dry sclerophyll forests on sandy soils of the coast and tablelands of southeastern Qld, northern and central NSW and northeastern Tas. **Family** Tremandraceae.

Tetratheca juncea

Low, sprawling shrub with angular, flattened branches to 1 m long. **Leaves** Alternate, usually reduced to narrow triangular scales to 3 mm long, or narrow-elliptic to 2 cm long and 5 mm wide. **Flowers** Lilac-pink with a dark brown centre, open, 15-20 mm across with 4 spreading petals and 8 stamens, solitary or in pairs on stalks to 1 cm long in the upper leaf axils. **Fruits** Obovate flattened capsules, 6-8 mm long, often beaked. **Flowering** Winter and spring. **Habitat** Sandy soils in heaths and dry sclerophyll forests of the north and central coasts of NSW. **Family** Tremandraceae.

Tetratheca thymifolia
Black-eyed Susan

Low, straggling shrub to 1 m high with hairy branches. **Leaves** Opposite, alternate or in whorls of 3-6, ovate to elliptical, hairy, 2-20 mm long and 1-8 mm wide. **Flowers** Lilac-pink or white with a dark brown centre, open, 15-30 mm across with 4 spreading petals and 8 stamens, solitary on stalks 5-30 mm long in the upper leaf axils. **Fruits** Flattened capsules, 5-7 mm long and 4-6 mm wide, often beaked. **Flowering** Mainly in spring. **Habitat** Widespread in sandy soils in heaths and dry sclerophyll forests of the coast and tablelands in southeastern Qld, NSW and southeastern Vic. **Family** Tremandraceae.

Crowea exalata
Small Crowea

Low shrub to 1 m high with minute hairs on the slender branchlets. **Leaves** Alternate, linear to spathulate or obovate, thick, paler below, 1-5 cm long and 2-6 mm wide with an aniseed aroma when crushed. **Flowers** Bright pink to mauve, rarely white inside, often greenish outside, waxy, open, 10-24 mm across with 5 pointed petals and 10 stamens with bearded anther appendages, solitary in the upper leaf axils. **Fruits** Capsules about 7 mm long. **Flowering** sporadically year round. **Habitat** Rocky sites on sandy soils in dry sclerophyll forests of the coast and western slopes of NSW and Vic. **Family** Rutaceae.

Crowea saligna
Willow-leaved Crowea

Low shrub to 1.5 m high with angular branches. **Leaves** Alternate, narrow-elliptic to lanceolate, thick with a prominent midrib, glossy dark-green above, 3-8 cm long and 4-20 mm wide. **Flowers** Bright pink to purple or rarely white inside, often greenish outside, waxy, open, 25-40 mm across with 5 pointed petals and 10 stamens with bearded anther appendages, solitary in terminal leaf axils. **Fruits** Capsules about 7 mm long. **Flowering** Mainly in spring. **Habitat** Rocky, sandy, sheltered sites in dry sclerophyll forests of the coast and tablelands in southeastern Qld and central eastern NSW. **Family** Rutaceae.

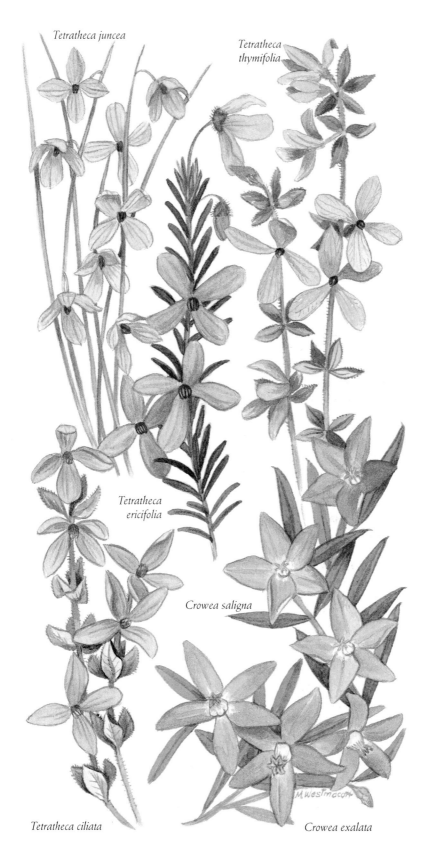

Tetratheca juncea

Tetratheca thymifolia

Tetratheca ericifolia

Crowea saligna

Tetratheca ciliata

Crowea exalata

M.Westmacott

Spergularia rubra
Sand Spurrey

Semi-erect or prostrate, annual or biennial herb to 20 cm high, covered with short hairs. **Leaves** Opposite, fleshy, linear to lanceolate, 3-25 mm long and 1-2 mm wide. **Flowers** White to pink, open, about 12 mm across with 5 pointed petals and 5 sepals about the same length, with 5-10 yellow stamens. They are arranged in loose terminal clusters. **Fruits** 3-valved capsules, 3-5 mm long with dark brown wingless seeds. **Flowering** Spring. **Habitat** Introduced from Europe, a common weed in disturbed sites, and often near watercourses and coastal marshes in all states. **Family** Caryophyllaceae.

Calandrinia remota
Round-leaved Parakeelya

Low, semi-prostrate annual or perennial herb with many flowering stems to 30 cm long. **Leaves** form a rosette around the base of the plant and on lower part of flowering stems stalkless, fleshy, broad linear to oblong or lanceolate, often red-pink, 3-11 cm long and 1-10 mm wide. **Flowers** Mauve-pink to purple, rarely white, with a whitish-yellow centre, open, about 25 mm across with 5 separate, spreading lobes, notched at the tips, and many yellow stamens, arranged in loose terminal racemes on long stalks. **Fruits** 3-valved ovoid capsules, 5-6 mm long with numerous red-brown seeds. **Flowering** Winter and spring. **Habitat** Red sand soils in arid shrublands of Qld, NSW, SA, WA and the NT. **Family** Portulacaceae.

Calytrix alpestris
Snow Myrtle

Erect wiry shrub to 3 m high with hairy branches. **Leaves** Alternate, aromatic, crowded, perpendicular to the branches, linear, covered with sparse stiff hairs, 2-5 mm long and about 5 mm wide. **Flowers** White to pale-pink, open, 7-15 mm across with 5 narrow lobes and numerous long, white, protruding stamens with red tips. They are arranged in small clusters in the upper leaf axils. **Fruits** Small nuts. **Flowering** Spring and early summer. **Habitat** Sandy soils in scrubs and open forests in Vic. and southeastern SA. **Family** Myrtaceae.

Calytrix fraseri
Purple Star Flower. Summer Fringe Myrtle

Small, spreading shrub to 1.5 m high. **Leaves** Alternate, aromatic, crowded, oblong to narrow-linear with curved-back margins, 3-5 mm long. **Flowers** Deep lilac to purple, open, 15-20 mm across with 5 narrow lobes and numerous long, protruding stamens. They are arranged in small clusters in the upper leaf axils. **Fruits** Small nuts about 15 mm long. **Flowering** Spring and early summer. **Habitat** Sandy coastal plains in partial shade in southwestern WA. **Family** Myrtaceae.

Calytrix tetragona
Common Fringe Myrtle

Erect or spreading shrub to 2 m high. **Leaves** Alternate, spirally arranged, often crowded and erect, linear to narrow-elliptic or oblong, 1-12 mm long and about 1 mm wide, thick and often hairy, often with finely-toothed margins. **Flowers** White to pink, open, 7-20 mm across with 5 narrow, pointed petals and numerous protruding stamens to 6 mm long. They are massed in leafy, terminal spikes. **Flowering** Mainly in spring and summer. **Habitat** Widespread heaths, dry sclerophyll forests and woodlands in rocky and sandy areas in all states except the NT. **Family** Myrtaceae.

Philotheca salsolifolia

Small twiggy shrub to 2 m high. **Leaves** Alternate, crowded, narrow-linear to needle-like, 4-120 mm long and 1-2 mm wide with small black glands on the leaf bases. **Flowers** Pink to lilac blue, open, 12-25 mm across with 5 spreading, pointed, hairy petals and 10 stamens united into a tube at their bases. They are terminal and solitary or 2-3 together, almost stalkless. **Fruits** Oblong, compressed capsules, 5-6 mm long. **Flowering** Spring and early summer. **Habitat** Widespread in rocky and sandy heaths, in forests and woodlands along the coast, table-lands and western slopes of southeastern Qld and NSW. **Family** Rutaceae.

Calytrix tetragona

Calytrix alpestris

Calytrix fraseri

Spergularia rubra

Philotheca salsolifolia

Calandrinia remota

Ricinocarpos bowmanii

Erect or spreading hairy shrub to 1.5 m high. **Leaves** Spirally arranged, linear to oblong with curved-back margins, rough, almost stalkless, 1-4 cm long and 1-3 mm wide. **Flowers** White to pink, open, 2-3 cm across with 4-6 spreading, separate lobes, male flowers have numerous stamens in a column, female flowers have red, wiry styles, both appear on the same plant, in terminal clusters of 3-6 males and one female. **Fruits** Globular furry capsules, 8-10 mm long. **Flowering** Winter, spring and early summer. **Habitat** Sandy and gravelly soils in dry sclerophyll forests and mallee communities in southeastern Qld and NSW. **Family** Euphorbiaceae.

Lasiopetalum behrii   Pink Velvet Bush

Slender erect shrub to 1.5 m high, covered with rusty hairs. **Leaves** Alternate, linear-oblong to narrow-elliptic with curved-back margins, covered with pale or rusty hairs below, 3-8 cm long and 5-20 mm wide. **Flowers** White to pink inside with pale-greenish-yellow markings and whitish outside, covered with downy hairs, open, 1-2 cm across with 5 spreading, separate lobes and 5 reddish stamens. They are arranged in small, drooping axillary clusters. **Fruits** Hairy capsules 5-7 mm long. **Flowering** Spring and early summer. **Habitat** Sand ridges in dry inland areas and mallee scrub in Vic. and SA. **Family** Sterculiaceae.

Lasiopetalum ferrugineum   Rusty Velvet Bush

Variable erect or prostrate shrub to 3 m high with rusty hairs on young branches. **Leaves** Alternate, narrow-elliptic or ovate to lanceolate, dull-green above, pale with rusty hairs below, 2-12 cm long and 4-40 mm wide. **Flowers** Whitish, hairy inside, covered with rusty hairs outside, open, 5-15 mm across with 5 separate, spreading lobes and 5 reddish stamens. They are arranged in small, dense, drooping axillary clusters. **Fruits** Hairy capsules, 3-5 mm across. **Flowering** Spring. **Habitat** Widespread in heaths and dry sclerophyll forests along the coast and adjacent ranges of southeastern Qld, NSW southeastern Vic. **Family** Sterculiaceae.

Lasiopetalum macrophyllum   Shrubby Velvet Bush

Erect, wiry shrub to 2 m high, with dense rusty hairs on the branches. **Leaves** Alternate, ovate to lanceolate, dull-green above, pale and dotted with rusty hairs below, 4-12 cm long and 1-4 cm wide. **Flowers** White to pink inside, whitish outside and covered with rusty hairs, open, 12-16 mm across with 5 separate, spreading lobes and 5 reddish stamens. They are arranged in small, crowded axillary clusters. **Fruits** Hairy capsules, 3-5 mm across. **Flowering** Spring. **Habitat** Widespread in sandy and rocky sites in forests and woodlands in southeastern Qld, NSW, Vic. and Tas. **Family** Sterculiaceae.

Rhytidosporum procumbens (syn. Billardiera procumbens)

Prostrate to erect shrub to 25 cm high. **Leaves** Alternate, scattered or clustered, stalkless, linear to narrow-oblong with flat or curled-under margins, 4-20 mm long and 1-3 mm wide. **Flowers** White, often tinged with mauve outside, open, 6-15 mm across with 5 separate spreading lobes and 5 stamens, terminal and solitary or 2-3 together on short terminal or subterminal stalks 2-3 mm long. **Fruits** Compressed, ovoid, leathery capsules 4-6 mm long and 5-8 mm wide. **Flowering** Spring and summer. **Habitat** Widespread in heaths and sclerophyll forests of the coast and tablelands in southeastern Qld, NSW, Vic., Kangaroo Is. in SA and Tas. **Family** Pittosporaceae.

Zygophyllum apiculatum   Pointed or Common Twin Leaf

Low shrub with a woody base to 40 cm high. **Leaves** Opposite, fleshy, divided into 2 ovate leaflets 15-40 mm long. **Flowers** Bright-yellow, open, 15-25 mm across with 5 separate, spreading petals and 10 stamens, solitary or in small axillary clusters. **Fruits** 5-angled capsules, 7-10 mm long. **Flowering** Winter and spring. **Habitat** Frequent ground cover in mallee scrubs and woodlands in drier areas of all mainland states. **Family** Zygophyllaceae.

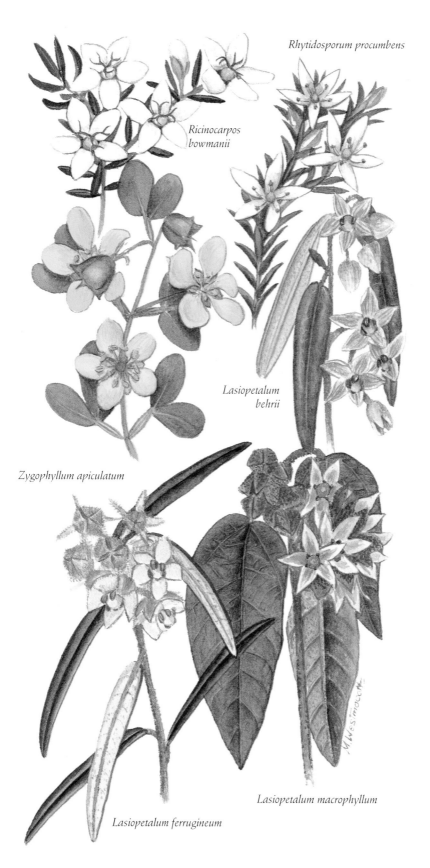

Rhytidosporum procumbens

Ricinocarpos bowmanii

Lasiopetalum behrii

Zygophyllum apiculatum

Lasiopetalum ferrugineum

Lasiopetalum macrophyllum

Muehlenbeckia florulenta (syn. M. cunninghamii)   Tangled Lignum

Erect, rounded shrub to 3 m high with densely entangling stiff, grey-green, striated, rigid branches, often ending in a spine. **Leaves** Linear to narrow-lanceolate, 15-70 mm long and 2-10 mm wide, falling early. **Flowers** Greenish-white to cream, open, 8-10 mm across with 5 separate, spreading lobes and 8 stamens, clustered in interrupted racemes along the stems, 2-12 cm long. **Fruits** Ovoid to conical shiny brown nuts, 3-4 mm long. **Flowering** Most of the year. **Habitat** Swampy sites and flood-prone areas in heavy soils, mainly inland in all mainland states. **Family** Polygonaceae.

Hibbertia dentata   Twining Guinea Flower

Prostrate or climbing shrub with wiry stems to 2 m long. **Leaves** Oblong-elliptic to ovate with sharply toothed or lobed margins, 4-7 cm long and 15-30 mm wide. **Flowers** Yellow, open, 25-50 mm across with 5 separate, spreading petals (rarely notched) and 5 pointed sepals, terminal and solitary. **Flowering** Spring and summer. **Habitat** Forests and rainforest margins of the coast and tablelands of southeastern Qld, NSW and southeastern Vic. **Family** Dilleniaceae.

Marsdenia suaveolens   Sweet-scented Doubah. Sweet Marsdenia

Variable semi-erect, trailing or climbing shrub with short hairs on the stems and milky latex. **Leaves** Opposite, oblong to lanceolate or ovate with curled-back margins, 2-7 cm long and 6-25 mm wide, dark-green above and paler below. **Flowers** White to greenish-white, open, 4-5 mm across with a short tube, bearded inside with 5 separate, spreading lobes. They are arranged compact axillary clusters. **Fruits** Narrow follicles, 5-10 cm long and 8-14 mm wide. **Flowering** Summer. **Habitat** Sandy heaths, open forests and rainforest gullies of the coast and ranges of NSW. **Family** Asclepiadaceae.

Tristania neriifolia   Dwarf Water Gum

Erect shrub to 3 m high. **Leaves** Opposite, narrow-lanceolate to narrow-elliptic, 35-90 mm long and 5-15 mm wide with a prominent central vein, dotted with oil glands. **Flowers** Yellow, open, 10-20 mm across with 5 separate, spreading lobes and numerous stamens grouped in 3-6 bundles. They are arranged in loose terminal or axillary clusters. **Fruits** 3-valved, thin-walled capsules, 4-5 mm diameter. **Flowering** Summer. **Habitat** Along watercourses, widespread in the central coast and ranges of NSW. **Family** Myrtaceae.

Asterolasia correifolia

Erect shrub to 2 m high with densely hairy stems. **Leaves** Alternate, elliptic to lanceolate, covered with soft pale hairs below, 3-9 cm long and 8-30 mm wide. **Flowers** White or cream, open, 10-12 mm across with 5 separate, spreading lobes covered with dense hairs outside, and 10 large yellow-tipped stamens. They are solitary or in small terminal or axillary clusters on stalks 3-15 mm long. **Fruits** Beaked, hairy capsules. **Flowering** Spring. **Habitat** Moist sandy gullies in wet sclerophyll forests of the coastal ranges in central eastern Qld and eastern NSW. **Family** Rutaceae.

Phebalium bullatum   Silvery Pheblium

Erect, wiry shrub to 2 m high covered with a whitish or rusty-brown scurf of minute circular scales. **Leaves** Alternate, narrow-oblong to wedge-shaped, thick with bubble-like projections, particularly on the margins, and a depressed midrib, 5-15 mm long and 1-3 mm wide. **Flowers** Yellow, open, 7-10 mm across with 5 separate, spreading lobes and 10 long, protruding stamens, arranged in terminal clusters of 1-8 flowers. **Flowering** Winter and spring. **Habitat** Sandy sites in mallee scrub and heaths in western Vic. and SA. **Family** Rutaceae.

Muehlenbeckia florulenta

Phebalium bullatum

Tristania neriifolia

Marsdenia suaveolens

Asterolasia correifolia

Hibbertia dentata

M.Westmacott

Phebalium bilobum
Notched Phebalium
Variable erect or sprawling shrub to 4 m high. **Leaves** Alternate, oblong or lanceolate with a notched tip and sometimes with toothed and curled-back margins, aromatic, paler below; thick, 7-15 mm long and 2-5 mm wide in the small form; thinner, 2-5 cm long and 3-9 mm wide in the larger form. **Flowers** White to reddish, open, 7-10 mm across with 5 separate, spreading lobes and 10 long, protruding stamens, arranged in small terminal clusters. **Flowering** Winter and spring. **Habitat** Wet sites in rocky outcrops and forests in Vic. and northeastern Tas. **Family** Rutaceae.

Phebalium squameum
Satinwood
Tall shrub or small tree, 3-12 m high, with scurfy scales on the branches. **Leaves** Alternate, oblong to lanceolate or narrow-elliptic with a pronounced midrib, silvery white below, dotted with glands above, 2-10 cm long and 5-22 mm wide. **Flowers** White to yellowish, rarely pink, open, 8-10 mm across with 5 separate, spreading lobes and 10 long, protruding stamens, arranged in axillary clusters of 3-20 flowers. **Flowering** Spring. **Habitat** Wet, sandy sites in forests and gullies in coastal southern Qld, eastern NSW, Vic. and Tas. **Family** Rutaceae.

Phebalium squamulosum
Forest or Scaly Phebalium
Variable tall or spreading shrub to slender tree, 1-7 m high, with rusty scales on the branches. **Leaves** Alternate, linear to oblong o elliptic with a pronounced midrib, silvery to reddish brown and scaly below, dotted with glands above, 1-7 cm long and 1-8 mm wide. **Flowers** Cream to yellowish, rarely pink, with rusty or silvery scales outside, open, 5-10 mm across with 5 separate, spreading lobes and 10 long, protruding stamens, arranged in terminal clusters of 4-8 flowers. **Fruits** Capsules about 3.5 mm long. **Flowering** Mainly in spring. **Habitat** Widespread in sandstone sites in forests and heaths of the coast and tablelands in Qld, NSW and eastern Vic. **Family** Rutaceae.

Philotheca buxifolia
Box-leaf Wax Flower
(syn Eriostemon buxifolius)
Low, compact shrub to 1.3 m high with rigid downy, ribbed branches. **Leaves** Alternate, elliptical to ovate or heart-shaped with a depressed midrib, thick, dark-green with small glands at the leaf bases, 6-18 mm long and 3-12 mm wide. **Flowers** Pale-pink or white inside, deeper pink outside, waxy, open, 15-30 mm across with 5 separate, spreading, hairy petals and 10 stamens, solitary in the leaf axils. **Flowering** Winter and spring. **Habitat** Sandy heaths of the central and south coasts NSW. **Family** Rutaceae.

Philotheca myoporoides
Long-leaf Wax Flower. Native Daphne
(syn Eriostemon myoporoides)
Variable tall or rounded shrub to 2 m high with aromatic glands on the leaves and branches. **Leaves** Alternate, stalkless, narrow-elliptical to obovate or lanceolate with a prominent midrib, leathery with small glands at the leaf bases, 15-110 mm long and 2-60 mm wide. **Flowers** White with pink buds, waxy, open, 15-20 mm across with 5 separate, spreading, hairy petals and 10 stamens, in axillary clusters of 1-8 flowers. **Fruits** re beaked capsules about 1 cm long. **Flowering** Mainly in spring and autumn. **Habitat** Widespread on rocky hillsides in heaths and dry sclerophyll forests in Qld, NSW and eastern Vic. **Family** Rutaceae.

Philotheca verrucosa
Fairy or Bendigo Wax Flower
(syn. Eriostemon verrucosus)
Low bushy shrub to 1 m high with glandular leaves and branches. **Leaves** Alternate, concave, ovate to heart-shaped, broader towards the tip, thick with small glands at the leaf bases, 6-15 mm long and 4-8 mm wide. **Flowers** Pink to white with pink buds, waxy, open, 1-2 cm across with 5 separate, spreading, hairy petals and 10 stamens, solitary or 2-3 together in the leaf axils. **Flowering** Winter and early spring. **Habitat** Stony, dry sites in open areas of Vic., southeastern SA and Tas. **Family** Rutaceae.

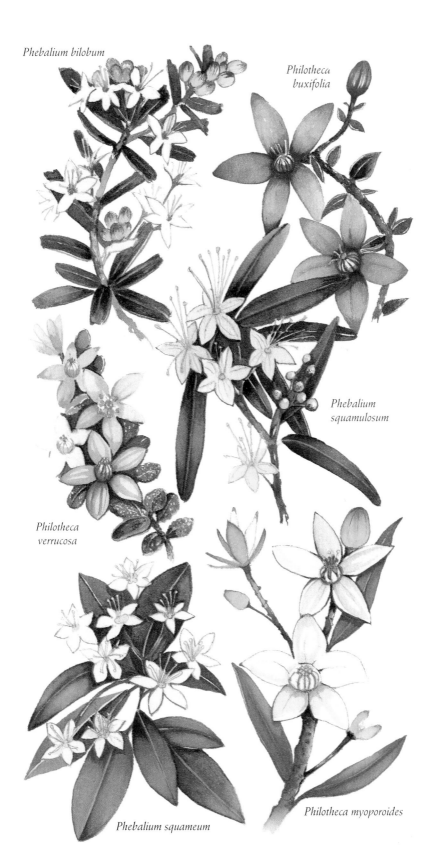

Phebalium bilobum

Philotheca buxifolia

Phebalium squamulosum

Philotheca verrucosa

Phebalium squameum

Philotheca myoporoides

Eriostemon australasius
Pink Wax Flower

Erect bushy shrub to 2 m high with minute hairs on the branches. **Leaves** Alternate, concave, linear to lanceo-late or narrow-elliptic, thick and leathery, 15-80 cm long and 2-14 mm wide. **Flowers** Pale-pink to white or red, waxy, open, 2-4 cm across with 5 separate, spreading lobes and 10 stamens, solitary in the upper leaf axils. **Fruits** Capsules about 9 mm long. **Flowering** Spring. **Habitat** Dry sclerophyll forests and heaths on sand-stones, mainly along the coast from southeastern Qld to just south of Sydney in NSW. **Family** Rutaceae.

Philotheca myoporoides leichhardtii *(syn. Eriostemon glasshousiensis)*

Erect bushy shrub to 2 m high with glandular branches. **Leaves** Alternate, stalkless, ovate to wedge-shaped, thick and leathery, with small glands at the leaf bases, 2-4 cm long and 6-8 mm wide. **Flowers** White to pink, waxy, open, about 2 cm across with 5 separate, spreading, hairy petals and 10 stamens, arranged in axillary clus-ters of 1-6 flowers. **Flowering** Spring. **Habitat** The Glasshouse Mountains and Girraween area of southeast-ern Qld. **Family** Rutaceae.

Philotheca obovalis *(syn. Eriostemon obovalis)*

Low shrub to 1 m high, with conspicuous glands on the leaves and branches. **Leaves** Alternate, ovate to spathu-late with a notched tip, crowded, thick and leathery, grey-green, with small glands at the leaf bases, 5-20 mm long and 4-6 mm wide. **Flowers** Pale-pink to white, waxy, open, 10-16 across with 5 separate, spreading, hairy petals and 10 stamens, solitary in the leaf axils on stalks 2-3 mm long. **Fruits** Capsules about 5 mm long. **Flowering** Mainly in spring. **Habitat** Sandstone ridges in heaths and dry sclerophyll forests, confined to the Blue Mtns in NSW. **Family** Rutaceae.

Philotheca spicata *(syn. Eriostemon spicatus)*
Pepper and Salt

Low shrub to 1 m high, with spindly, usually hairy branches. **Leaves** Alternate, stalkless, narrow-linear to very nar-row-elliptic, covered with small glands, 6-20 mm long. **Flowers** Pink to white or bluish, waxy, open, 6-12 mm across with 5 separate, spreading lobes and 10 stamens, arranged in a terminal raceme to 20 cm long. **Flowering** Winter and spring. **Habitat** Sandy or gravelly soils of coastal southwestern WA. **Family** Rutaceae.

Cheiranthera cyanea *(syn. C. linearis)*
Finger Flower

Low, slender shrub to 50 cm high. **Leaves** Alternate, often clustered, stalkless, narrow-linear, sharply-pointed, sometimes with minutely toothed curled-back margins, 1-6 cm long and 1-4 mm wide. **Flowers** Pale to deep blue, open, 2-4 cm across with 5 separate, spreading lobes and 5 large yellow stamens, finger-like on one side of the flower, terminal and solitary or 2-5 together on long, slender stalks. **Fruits** Hard, brown, oblong, flattened capsules, 12-18 mm long. **Flowering** Summer and autumn. **Habitat** Sandy soils in sclerophyll forests and woodlands mainly in the slopes and tablelands of southeastern Qld, NSW and Vic. **Family** Pittosporaceae.

Ricinocarpos pinifolius
Wedding Bush

Erect to spreading shrub to 3 m high. **Leaves** Spirally arranged, linear with curled-under margins, paler below, 1-4 cm long and 1-3 mm wide. **Flowers** (male) White, open, 2-3 cm across with 4-6 separate, spreading lobes and numerous yellow stamens in a column, arranged in terminal clusters of 3-6 male flowers around a red, globular, female flower about 6 mm across. **Fruits** Globular, densely-spiny capsules about 12 mm across. **Flowering** Winter and spring. **Habitat** Sandy soils in heaths and open forests of coastal Qld, NSW, Vic. and northeastern Tas. **Family** Euphorbiaceae.

Philotheca spicata

*Philotheca
myoporoides
leichhardtii*

*Eriostemon
australasius*

Philotheca obovalis

Ricinocarpos pinifolius

Cheiranthera cyanea

Leptospermum polygalifolium (syn. L. flavescens)   Yellow Tea Tree

Erect shrub or small tree, 1-7 m high with several stems and firm, smooth bark. **Leaves** Alternate, narrow, oblong to elliptic, rigid, flat or with slightly curled-back margins, 5-20 mm long and 1-5 mm wide on very short stalks, slightly lemon-scented. **Flowers** White to yellowish, solitary, open, 5-16 mm across with 5 separate spreading lobes around a green central disc surrounded by numerous stamens. **Fruits** Woody, 5-valved, domed capsules, 5-10 mm diameter. **Flowering** Spring and summer. **Habitat** Damp sites in sandy soils in heaths and woodlands of the coast and ranges in Qld, and NSW. **Family** Myrtaceae.

Leptospermum grandifolium   Woolly or Mountain Tea Tree

Erect dense shrub or small tree, 1-6 m high, with smooth, exfoliating bark. **Leaves** Alternate, oblong, ovate to narrow-elliptic or obovate, pointed, 10-35 mm long and 3-10 mm wide, shiny-green or reddish green above, paler and silky-hairy below. **Flowers** White, solitary and stalkless, open, to 18 mm across with 5 separate spreading lobes around a dark central disc surrounded by numerous stamens and 5 red sepals. **Fruits** Domed, 5-valved, woody capsules 8-10 mm diameter. **Flowering** Spring and summer. **Habitat** sandy swamps and rocky sites by cool mountain streams along the coast and ranges of central eastern and southeastern NSW and southeastern Vic. **Family** Myrtaceae.

Leptospermum juniperinum   Prickly Tea Tree

Erect rigid shrub, 2-3 m high with smooth bark. **Leaves** Alternate, concave, rigid, linear to narrow-ovate or lanceolate, sharply-pointed, 5-18 mm long and 1-10 mm wide. **Flowers** White, open, 6-10 mm across with 5 separate spreading lobes around a green central disc surrounded by numerous stamens, usually solitary and stalkless. **Fruits** Domed, woody, 5-valved capsules 5-8 mm diameter. **Flowering** Mainly in spring and summer. **Habitat** Widespread on poorly-drained soils in heaths and woodlands along the coast of southeastern Qld and NSW. **Family** Myrtaceae.

Leptospermum myrsinoides   Heath or Silky Tea Tree

Erect wiry shrub, 1-2 m high with silky young stems. **Leaves** Alternate, rigid, concave, lanceolate, broadest towards the tip, 4-12 mm long and 1-3 mm wide, with incurved margins, often silky-hairy below. **Flowers** White or pink, solitary, open, 10-15 mm across with 5 separate spreading lobes around a greenish-brown central disc surrounded by numerous stamens inserted in a silky base. **Fruits** Cup-shaped, 4-5 valved, stalkless capsules, 4-6 mm across. **Flowering** Spring. **Habitat** Poor sandy soils in mallee and coastal heaths along the south coast of NSW, Vic. and southeastern SA. **Family** Myrtaceae.

Leptospermum nitidum   Shiny Tea Tree

Erect spreading shrub to 3 m high. **Leaves** Alternate, lanceolate, broadest towards the tip, sharply-pointed, 15-25 mm long and 5-7 mm wide, shiny dark-green. **Flowers** White, solitary, open, 2-3 cm across with 5 separate, spreading lobes around a green central disc surrounded by numerous stamens and 5 small red sepals. **Fruits** Domed, 5-valved, brown capsules to 9 mm diameter. **Flowering** Spring. **Habitat** Rocky sites in mountains and open forests of Tas. **Family** Myrtaceae.

Kunzea ericoides (syn. Leptospermum phylicoides)   Burgan

Erect dense shrub or small tree to 5 m high with soft hairs on young stems. **Leaves** Alternate, narrow-elliptic to lanceolate, pointed, 6-25 mm long and 1-5 mm wide. **Flowers** Creamy-white, open, 5-10 mm across with 5 separate, spreading lobes around a brown central disc surrounded by numerous stamens, 1-4 mm long. They are solitary or crowded in small leafy racemes. **Fruits** Woody, cup-shaped capsules, 2-4 mm long and 3-5 mm across. **Flowering** Spring and summer. **Habitat** Near creeks in heaths and open forests, common at higher elevations, along the coast and ranges of southeastern Qld, NSW and Vic. **Family** Myrtaceae.

Leptospermum myrsinoides

Leptospermum nitidum

Kunzea ericoides

Leptospermum grandifolium

Leptospermum polygalifolium

Leptospermum juniperinum

M. Westmacott

Leptospermum scoparium
Manuka. Broom Tea Tree

Erect, spreading shrub with smooth bark and silky young stems, to 2 m high, low or prostrate in alpine areas.
Leaves Alternate, concave, broad-lanceolate, 7-20 mm long and 2-10 mm wide, sharply-pointed. **Flowers** White, rarely pink or red, solitary, open, 8-20 mm across with 5 separate spreading lobes around a green central disc surrounded by numerous stamens, 2-4 mm long. **Fruits** Woody, domed, 5-valved capsules, 6-10 mm diameter.
Flowering Spring and summer. **Habitat** Rocky and sandy sites in heaths, often near fast-flowing streams along the coast and ranges of southern NSW, Vic. and Tas. **Family** Myrtaceae.

Leptospermum squarrosum
Peach Blossom Tea Tree

Erect bushy shrub, 1-4 m high with smooth bark and silky young stems. **Leaves** Alternate, concave, lanceolate to broad-elliptic or ovate, sharply-pointed, rigid, 5-15 mm long and 2-5 mm wide. **Flowers** White or pale-pink, solitary, open, 1-2 cm across with 5 separate, spreading lobes around a green central disc surrounded by numerous stamens, 3-4 mm long. **Fruits** Woody, cup-shaped, domed, 5-valved capsules, 8-12 mm diameter.
Flowering Mainly in autumn. **Habitat** Widespread in sclerophyll shrublands on sandy soils of the central and southern coast and tablelands of NSW. **Family** Myrtaceae.

Baeckea crassifolia
Desert Baeckea. Desert Heath Myrtle

Low slender shrub to 75 cm high. **Leaves** Opposite, thick, linear to narrow obovate, blunt tipped, 1-3 mm long and 0.5 mm wide. **Flowers** White to pale mauve, open, 4-8 mm across with 5 separate spreading lobes surrounding a central disc with 5-12 stamens. They are solitary in the upper leaf axils on short stalks. **Fruits** 3-valved woody capsules about 2 mm across. **Flowering** Winter and spring. **Habitat** Mallee scrub, sandhills and heaths in drier inland areas of NSW, southwestern Vic., SA and southwestern WA. **Family** Myrtaceae.

Baeckea gunniana
Alpine Baeckea

Erect or prostrate, aromatic, densely-branched shrub to 1 m high. **Leaves** Opposite, crowded, often concave with curled-back tips, obovate to narrow-oblong, 2-5 mm long and about 1 mm wide, thick, fleshy and dotted with oil glands. **Flowers** White, open, 4-6 mm across with 5 separate spreading lobes surrounding a central disc with 5-10 stamens. They are solitary in the upper leaf axils on short stalks. **Fruits** Capsules 1.5-2 mm diameter.
Flowering Summer. **Habitat** Wet sites above 500 m in heaths and woodlands of central and southern NSW, southeastern Vic. and Tas. **Family** Myrtaceae.

Baeckea ramosissima
Rosy Heath Myrtle. Rosy Baeckea

Erect or prostrate shrub to 60 cm high with spreading branches. **Leaves** Opposite, narrow-linear to narrow-ovate, 3-13 mm long and 1-3 mm wide, dotted with oil glands. **Flowers** Mauve to pink or white, open, 6-15 mm across with 5 separate, spreading lobes surrounding a central disc with 10 stamens. They are solitary on short axillary stalks. **Fruits** Capsules. **Flowering** Winter, spring and summer. **Habitat** Sandy heaths and open forests at various altitudes of eastern NSW, Vic., the Southern Lofty region of SA, Kangaroo Is. and Tas. **Family** Myrtaceae.

Baeckea virgata
Tall or Twiggy Baeckea

Erect or prostrate shrub to 4 m high. **Leaves** Opposite, stalkless, narrow-oblong to narrow-lanceolate or elliptic, 7-25 mm long and 1-7 mm wide, dotted with oil glands, dark-green above and paler below. **Flowers** White, rarely pink, open, 5-10 mm across with 5 separate, spreading lobes around a green central disc surrounded by 7-20 stamens. They are arranged in axillary clusters of 2-9 flowers. **Fruits** 3-valved, cup-shaped capsules, 2-4 mm across.
Flowering Late spring and summer. **Habitat** Heaths and sclerophyll forests, often along watercourses along the coast and ranges of Qld, NSW, southeastern Vic. and the Victoria River district of the NT. **Family** Myrtaceae.

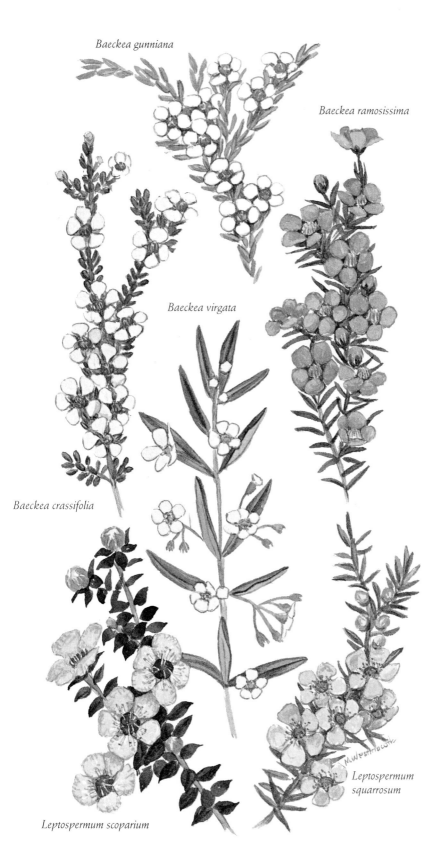

Baeckea gunniana

Baeckea ramosissima

Baeckea virgata

Baeckea crassifolia

Leptospermum scoparium

Leptospermum squarrosum

Verticordia insignis

Erect slender shrub to 60 cm high. **Leaves** Opposite, broad, oval to oblong, 4-8 mm long. **Flowers** Pink with deep red centres, open, about 12 mm across with 5-9 fringed, orbicular, separate, spreading lobes, massed in loose, leafy terminal clusters on long, slender stalks. **Flowering** Spring. **Habitat** Sandy and gravelly sites in southwestern WA. **Family** Myrtaceae.

Drosera arcturi   Alpine Sundew

Upright, insectivorous, perennial herb to 10 cm high. **Leaves** Narrow-oblong, crowded or forming a rosette around the base of the plant, 2-7 cm long and 3-5 mm wide, fleshy, light green or bronze, covered above with sticky hairs which trap and close over small insects. **Flowers** White, open, 1-2 cm across with 5 separate, spreading lobes, mostly solitary and terminal on thick stems 3-10 cm long. **Fruits** Small ovoid capsules. **Flowering** Spring and summer. **Habitat** Wet, mossy sites in moorlands above 1200 m in the southern tablelands of NSW, southeastern Vic. and Tas. **Family** Droseraceae.

Drosera auriculata   Tall Sundew

Erect, slender, insectivorous, perennial herb to 20 cm high. **Leaves** (lower) In a flat rosette around the base, circular to shield-shaped, 2-6 mm diameter on slender stalks attached to their centres and covered above with sticky hairs which trap and close over small insects; stem leaves are alternate or clustered with 2 pointed lobes, 4-6 mm across. **Flowers** White or pale-pink, open, 10-15 mm across, with 5 rounded and sometimes notched, separate, spreading lobes, arranged in loose terminal clusters, 3-10 cm long of 2-8 flowers. **Fruits** Small capsules. **Flowering** Spring and summer. **Habitat** Widespread in moist sites in Qld, NSW, Vic., SA and Tas. **Family** Droseraceae.

Drosera binata   Forked Sundew

Erect, slender, insectivorous, perennial herb with flowering stems to 50 cm high. **Leaves** Linear, fleshy, 2-15 cm long on stalks 4-30 cm long. They are forked into 2, 4 or 8 segments, pale-green to reddish, arising from the base of the stem and covered with sticky hairs which trap and close over small insects. **Flowers** White or pink, open, 10-25 mm across, with 5 rounded and sometimes notched separate, spreading lobes, terminal on tall, branched stems of 15-30 flowers. **Fruits** Small capsules. **Flowering** Spring and summer. **Habitat** Wet, peaty soils in shady sites of the coast and tablelands in southeastern Qld, NSW, Vic., southeastern SA and Tas. **Family** Droseraceae.

Drosera whittakeri   Scented Sundew

Low, perennial, insectivorous herb to 8 cm high. **Leaves** Broad spathulate, forming a flat rosette around the base of the plant, 10-35 mm long and 5-15 mm wide, fleshy, light green to bronze, covered above with sticky hairs which trap and close over small insects. **Flowers** White, fragrant, 20-25 mm across with 5 rounded and sometimes notched separate, spreading lobes, solitary and terminal on a thick flowering stem 2-4 cm long. **Fruits** Small capsules. **Flowering** Winter and spring. **Habitat** Widespread in damp, open sites in Vic. and southeastern SA. **Family** Droseraceae.

Astartea fascicularis

Spreading shrub to 3 m high. **Leaves** Opposite, clustered along the branches, narrow-linear, thick and flat or triangular in section, aromatic when crushed, to 5 cm long. **Flowers** White to pale-pink with a green to red central disc, open, about 1 cm across with 5 orbicular, separate, spreading lobes, solitary and profuse in the leaf axils on stalks about 3 mm long. **Fruits** 3-celled flattened capsules. **Flowering** Most of the year. **Habitat** Damp, sandy soils in southwestern WA. **Family** Myrtaceae.

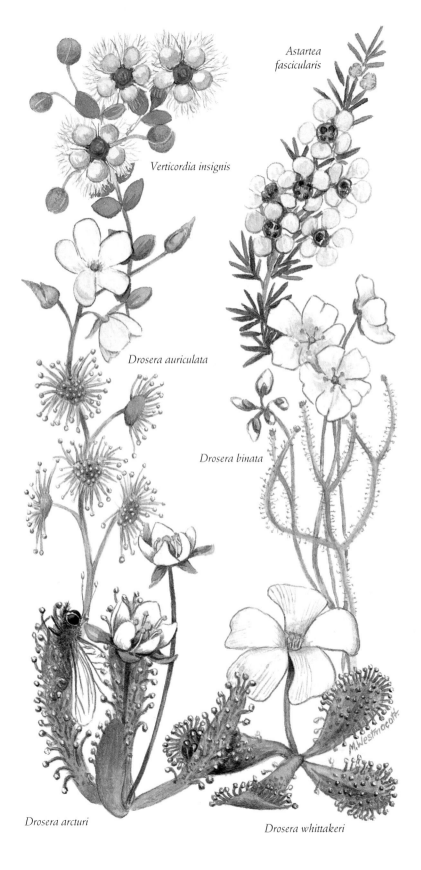

Astartea
fascicularis

Verticordia insignis

Drosera auriculata

Drosera binata

Drosera arcturi

Drosera whittakeri

M.Westmacott

Pomaderris andromedifolia

Erect shrub to 2 m high with stiff branches and rusty hairs on the young stems. **Leaves** Alternate, variable, narrow-elliptic to lanceolate or rarely oblanceolate, with a depressed midrib, covered with pale-yellow or rusty silky hairs below, 6-50 mm long and 3-15 mm wide. **Flowers** Golden yellow to cream, silky-hairy outside, open, about 5 mm across with 5 separate spreading lobes and long yellow stamens, arranged in compact terminal clusters 1-4 cm across. **Fruits** Capsules with long, whitish hairs. **Flowering** Mainly in spring. **Habitat** Mostly near streams of the coast and ranges of southeastern Qld, NSW and southeastern Vic. **Family** Rhamnaceae.

Pomaderris ferruginea   Rusty Pomaderris

Erect, open shrub to 4 m high with soft, rusty hairs on the stems. **Leaves** Alternate, oblong-elliptic to broad-lanceolate, dull dark-green above with distinct veins and rusty woolly hairs below, 3-10 cm long and 10-35 mm wide. **Flowers** Yellow to white, silky-hairy outside, open, about 5 mm across with 5 separate spreading lobes and 5 long stamens, arranged in compact terminal or axillary clusters, 5-10 cm across. **Fruits** 3-valved, ellipsoid, hairy capsules about 4 mm long. **Flowering** Spring. **Habitat** Mostly near streams in open forests of the coast and ranges of southeastern Qld, NSW, central and eastern Vic. **Family** Rhamnaceae.

Pomaderris lanigera   Woolly Pomaderris

Erect shrub or small tree, 1-5 m high with soft, rusty hairs on the stems. **Leaves** Alternate, lanceolate to narrow-ovate or elliptical, usually covered with soft hairs on both sides, paler or rusty below, 2.5-12 cm long and 12-40 mm wide. **Flowers** Golden yellow, silky-hairy outside, open, 6-10 mm across with 5 separate spreading lobes and 5 long stamens, arranged in terminal, hairy clusters, 3-12 cm across. **Fruits** 3-valved ellipsoid capsules about 3 mm long. **Flowering** Spring. **Habitat** Widespread on sandy sites in open forests of the coast and ranges of southeastern Qld, NSW and Vic. **Family** Rhamnaceae.

Ranunculus inundatus   River Buttercup

Weak, upright perennial herb to 35 cm high. **Leaves** Alternate or arising from the base of the plant, on stalks to 15 cm long, divided into usually 3 narrow-linear segments, 1-3 cm long and 1-2 mm wide. **Flowers** Shiny yellow, open, 10-15 mm across with 5-9 separate spreading petals and numerous yellow stamens surrounding a central disc. They are solitary and terminal on long axillary or leaf-opposing stalks, 7-35 cm long. **Fruits** Ridged obovate achenes, 1-2 mm across with a small beak. **Flowering** Spring and summer. **Habitat** Widespread in damp areas subject to flooding, sometimes aquatic, in southeastern Qld, NSW, Vic., southeastern SA and northeastern Tas. **Family** Ranunculaceae.

Ranunculus millanii   Dwarf Buttercup

Low, spreading, perennial herb with stems to 8 cm long. **Leaves** arise from the base of the plant, divided into 3-5 narrow-linear segments, 1-5 cm long. **Flowers** White to cream, open, 6-15 mm across with 5 (rarely to 12) separate spreading petals and numerous yellow stamens surrounding a central disc. They are solitary and terminal on short axillary stalks. **Fruits** Achenes with an erect or incurved beak. **Flowering** Summer. **Habitat** Alpine and sub-alpine areas subject to flooding in the southern tablelands of NSW and southeastern Vic. **Family** Ranunculaceae.

Oxalis corniculata   Creeping Oxalis. Yellow Wood Sorrel

Perennial herb with ascending or creeping branches, to 30 cm long, covered with hairs. **Leaves** Palmately divided usually into 3 heart-shaped leaflets, folded downwards, 4-18 mm long and 8-23 mm wide. **Flowers** Yellow, open, 12-15 mm across with 5 separate spreading petals and 10 stamens. They are solitary or in clusters of 2-6 flowers on long, slender, axillary stalks. **Fruits** Cylindrical, hairy capsules, 6-25 mm long and 1-4 mm wide. **Flowering** Spring, summer and autumn. **Habitat** Imported from Europe, now widespread in many situations throughout Australia. **Family** Oxalidaceae.

Ranunculus inundatus

Oxalis corniculata

Ranunculus millanii

Pomaderris lanigera

Pomaderris andromedifolia

Pomaderris ferruginea

Erodium crinitum
Blue Storksbill. Blue Crowfoot

Low annual herb to 50 cm high with scattered long white hairs on the stems. **Leaves** arise from the base of the plant, they are deeply dissected into 3 ovate, toothed lobes 1-4 cm long and 1-3 cm wide, with stalks 2-13 cm long. **Flowers** Blue with white or yellowish veins towards the base, open, 7-15 mm across with 5 separate, spreading lobes, arranged in terminal clusters of 2-6 flowers. **Fruits** Spirally twisted, sharply-pointed, hairy seeds, 4-7 cm long. **Flowering** From July to April. **Habitat** Widespread in sandy soils in open woodlands, saltbush communities and grasslands of all mainland states. **Family** Geraniaceae.

Pelargonium australe
Native Storksbill

Erect to semi-prostrate, downy, perennial herb to 60 cm high. **Leaves** Opposite, ovate, entire or deeply dissected into 5-7 toothed lobes, soft and hairy, 2-9 cm long and 2-8 cm wide on stalks to 15 cm long. **Flowers** Pink to white with purplish veins, open, about 16 mm across with 5 separate, spreading, unequal lobes, arranged in terminal clusters of 4-12 flowers. **Fruits** Hairy, plumed pods, 8-15 mm long with a single seed. **Flowering** Spring and summer. **Habitat** Sand dunes, coastal cliffs and rocky inland outcrops of southern Qld, NSW, Vic., SA, southwestern WA and Tas. **Family** Geraniaceae.

Pelargonium rodneyanum
Magenta Storksbill

Upright, softly hairy perennial herb to 40 cm high. **Leaves** Opposite, mostly arising from the base of the plant, ovate, 1-5 cm long and 15-40 mm wide on stalks to 7 cm long, often dissected into 5-7 shallow lobes with wavy margins and minute hairs. **Flowers** Reddish purple to deep-pink with deep purple streaks, open, 3-4 cm across with 5 long, separate, spreading, unequal lobes and 10 stamens in 2 rows. They are arranged in terminal clusters of 2-7 flowers on stalks 5-12 cm long. **Fruits** Hairy, plumed seeds, 18-22 mm long. **Flowering** From spring to autumn. **Habitat** Exposed rocky outcrops in sclerophyll forests of the central and southern tablelands of NSW, Vic. and southeastern SA. **Family** Geraniaceae.

Neopaxia australasica (syn. Montia australasica)
White Purslane

Prostrate, creeping perennial herb with stems to 30 cm long. **Leaves** Alternate, fleshy, sometimes floating, narrow-linear to lanceolate, broader towards the tip, 2-10 cm long and 1-4 mm wide. **Flowers** White or pale-pink, open, 1-2 cm across with 5 separate, spreading petals and 5 stamens united into a short tube. They are arranged in terminal racemes of 1-4 flowers. **Fruits** 3-valved globular capsules 2-4 mm across with black shiny seeds. **Flowering** Spring and summer. **Habitat** Freshwater swamps and stream banks from sea-level to the alps, where it may form dense mats, in NSW and Vic. **Family** Portulacaceae.

Calandrinia balonensis
Broad-leaved Parakeelya

Prostrate annual herb with erect flowering stems to 30 cm long. **Leaves** Alternate or arising from the base of the plant, fleshy, linear or broad-lanceolate, 2-12 cm long and 5-20 mm wide. **Flowers** Mauve-pink to purple, rarely white, with a yellow centre, open, 25-30 mm across with 5 separate, spreading lobes, notched at the tips, and 20-80 yellow stamens. They are arranged in loose terminal racemes of 3-4 flowers on stalks 2-3 cm long. **Fruits** Ovoid 3-valved capsules, 7-9 mm long with numerous small black or dark-red seeds. **Flowering** Winter and spring. **Habitat** Sandy soils in arid areas of all mainland states except Vic. **Family** Portulacaceae.

Viola hederacea
Ivy-leaf Violet

Perennial herb with erect, hairy or smooth stems to 10 cm long. **Leaves** Alternate or arising from the base of the plant, kidney-shaped to almost orbicular with entire or coarsely-toothed margins, 1-20 mm long and 1-30 mm wide on stalks 2-6 cm long. **Flowers** Pale-violet to white, usually blotched with purple, open, 14-20 mm across with 5 spreading lobes, the lowest usually broader than the rest, and 5 stamens, solitary on axillary stalks to 10 cm or more long. **Fruits** 3-valved ovoid capsules, 4-6 mm long. **Flowering** Spring and summer. **Habitat** Widespread in damp, sheltered sites in forests and woodlands of the coast and tablelands in Qld, NSW, Vic., southeastern SA and Tas. **Family** Violaceae.

*Calandrinia
balonensis*

Pelargonium rodneyanum

Pelargonium australe

*Viola
hederacea*

Erodium crinitum

Neopaxia australasica

M. Westmacott.

Geranium solanderi
Native Geranium

Sprawling, coarsely hairy, perennial herb with stems to 50 cm long. **Leaves** Opposite, circular, 1-3 cm long and 1.5-5 cm wide, divided into 5-10 obovate lobes, each divided into 3-5 lobes or teeth. **Flowers** Pale-pink, often with yellow veins, open, 12-24 mm across with 5 spreading, notched, overlapping lobes and 10 stamens, solitary or in pairs on hairy axillary stalks, 1-4 cm long. **Fruits** Dry, hairy seed pods, 12-25 mm long each with a black seed. **Flowering** Spring and summer. **Habitat** Widespread in grassy areas and open forests in Qld, NSW, Vic., SA, southwestern WA and Tas. **Family** Geraniaceae.

Geranium antrorsum

Compact, trailing, hairy, perennial herb to 20 cm high. **Leaves** Opposite, crowded, arising from the base of the plant, circular, 1-4 cm long and 15-30 mm wide, palmately divided into 5-7 rounded lobes with notched margins, dark-green with stiff hairs, further divided into 3 smaller lobes towards the tips. **Flowers** Pale to deep pink, striped with dark pink, open, 12-24 mm across with 5 spreading, overlapping lobes and 10 stamens, solitary on hairy, slender stalks, 1-4 cm long. **Fruits** Pods with spreading hairs, 13-18 mm long splitting to reveal a single greenish-black seed. **Flowering** Summer. **Habitat** Alpine herbfields, woodlands and tussock grasslands at higher altitudes in the southern tablelands of NSW and southeastern Vic. **Family** Geraniaceae.

Linum marginale
Native or Wild Flax

Upright perennial herb to 60 cm high. **Leaves** Alternate, narrow-linear to narrow-elliptic, stalkless, 5-20 mm long and 1-3 mm wide. **Flowers** Blue, open, 1-2 cm across with 5 overlapping, rounded lobes and a protruding column of fused stamens, arranged in loose, terminal clusters on long stalks. **Fruits** Globular capsules, 4-6 mm across with brown seeds about 3 mm long. **Flowering** Spring and summer. **Habitat** Widespread in woodlands, open forests and swamp margins in all states except NT. **Family** Linaceae.

Ranunculus lappaceus
Common Buttercup

Upright perennial herb to 50 cm high. **Leaves** Very variable, alternate or arising from the base of the plant, hairy, ovate to triangular, 12-80 mm long and wide, divided into usually 3 ovate or wedge-shaped segments with entire or lobed margins. **Flowers** Shiny yellow, 15-40 mm across usually with 5 overlapping, spreading petals and numerous yellow stamens surrounding a central disc. They are solitary and terminal on long axillary or leaf-opposing stalks, 15-50 cm high. **Fruits** Obovate achenes, 1-4 mm long with an erect, hooked beak, 1-2 mm long. **Flowering** Spring and summer. **Habitat** Widespread in damp areas subject to flooding, in grasslands and forests to 1500 m in Qld, NSW, Vic., SA and northeastern Tas. **Family** Ranunculaceae.

Keraudrenia hillii

Erect shrub to 3 m high, covered with dense, rusty hairs. **Leaves** Alternate, linear to lanceolate, leathery, dark-green above and covered with rusty or whitish hairs below, 4-11 cm long and 4-10 mm wide. **Flowers** Blue to purple, open, 10-18 mm across with 5 overlapping lobes with prominent midribs and 5 yellow stamens, solitary or in terminal clusters of 2-5 flowers on hairy stalks. **Fruits** Hairy capsules, 12-18 mm across. **Flowering** Spring and summer. **Habitat** Sclerophyll forests of the coast and tablelands in northern NSW and southeastern Qld. **Family** Sterculiaceae.

Viola betonicifolia
Purple Violet

Upright perennial herb to 20 cm high. **Leaves** Alternate or arising from the base of the plant, oblong to lanceolate or ovate, sometimes with toothed margins, 1-6 cm long and 5-30 mm wide on stalks 2-10 cm long. **Flowers** Blue to violet, often with deep purple streaks, open, 1-2 cm across with 5 spreading, overlapping lobes, the lowest slightly shorter than the rest, and 5 stamens. Usually solitary on axillary stalks 5-20 cm long. **Fruits** Ellipsoid 3-valved capsules, 7-13 mm long. **Flowering** Summer. **Habitat** Widespread in open forests and woodlands in Qld, NSW, Vic., southeastern SA and Tas. **Family** Violaceae.

Ranunculus lappaceus

Keraudrenia hillii

Linum marginale

Viola betonicifolia

Geranium antrorsum

Geranium solanderi

Alyogyne hakeifolia
Red-centred Hibiscus

Erect branching shrub to 3 m high. **Leaves** Alternate, divided 1-3 times into linear segments each 5-10 cm long and 1-2 mm wide, cylindrical or grooved. **Flowers** Mauve with a dark-red centre, open, 5-10 cm across with 5 overlapping, rounded petals and a protruding column of fused, rust-coloured stamens, solitary and axillary. **Fruits** 5-celled capsules about 2 cm long. **Flowering** Spring and summer. **Habitat** Widespread in a variety of drier locations in SA and southwestern WA. **Family** Malvaceae.

Alyogyne huegelii
Lilac Hibiscus

Erect shrub to 3 m high with rigid, hairy branches. **Leaves** Variable, 2-7 cm long, deeply divided into 3-5 obovate to oblong, irregularly shaped lobes with small hairs on both sides. **Flowers** Lilac to pink or white, open, 7-12 cm across with 5 overlapping rounded petals each 5-8 cm long, a star-like yellow, protruding style, and a column of fused stamens, solitary, on long axillary stalks. **Fruits** Ovoid hairy capsules. **Flowering** Spring and summer. **Habitat** Rocky hillsides and gorges in SA and southwestern WA. **Family** Malvaceae.

Hibiscus diversifolius
Swamp or Yellow Hibiscus

Erect spreading shrub to 2 m high with prickly, hairy branches. **Leaves** Alternate, rough, very variable, oblong to broad-lanceolate or orbicular, 2-10 cm long and 2-11 cm wide on stalks to 13 cm long, often divided into 3-5 lobes with irregularly toothed margins. **Flowers** Yellow with a red or purplish centre, open, 7-13 cm across with 5 overlapping rounded petals and a protruding column of fused stamens, solitary or arranged in an axillary raceme. **Fruits** Hairy 5-celled capsules about 2 cm long. **Flowering** Most of the year. **Habitat** Wet and swampy sites in scrubs and rainforest margins along the coast of Qld and NSW. **Family** Malvaceae.

Hibiscus heterophyllus
Native Rosella

Tall, rather prickly, erect shrub or small tree to 6 m high. **Leaves** Alternate, linear to narrow-ovate, 5-20 cm long and 5-110 mm wide on stalks to 6 cm long, entire or deeply divided into 3 lobes with toothed margins, rough with prominent veins below. **Flowers** White, pinkish or yellow with a deep red to purple centre, open, 10-12 cm across with 5 overlapping rounded petals and a protruding column of fused stamens, solitary in the upper leaf axils. **Fruits** 5-celled ovoid, hairy, brown capsules, 15-20 mm long. **Flowering** Spring and summer. **Habitat** Rainforest margins, tall forests and river banks of the coast and tablelands in Qld, northern and central eastern NSW. **Family** Malvaceae.

Hibiscus splendens

Tall shrub or small tree to 7 m high with downy, prickly stems. **Leaves** Alternate, the lower ones are 6-20 cm long and deeply divided into 3-5 ovate to heart-shaped lobes with toothed margins, covered in dense woolly hairs. Upper leaves are lanceolate to ovate, 3-8 cm long. **Flowers** Rose-pink with a reddish stripe, open, about 15 cm across with 5 overlapping rounded petals and a protruding column of fused red stamens, solitary on long axillary stems. **Fruits** 5-celled ovoid, hairy capsules, 25-30 mm long. **Flowering** Winter, spring and summer. **Habitat** Moist areas along river banks and rainforest margins in eastern Qld, northern and central eastern NSW. **Family** Malvaceae.

Hibiscus tiliaceus
Coast Cottonwood. Tree Hibiscus

Tall bushy shrub or small dense tree to 8 m high. **Leaves** Alternate, heart-shaped with wavy margins, 5-20 cm long and 5-18 cm wide on stalks to 14 cm long, whitish hairy below. **Flowers** Yellow with a red to purple centre, open, about 8 cm across with 5 overlapping rounded petals and a protruding column of fused stamens, arranged in a terminal raceme. **Fruits** Oblong to globular, hairy, 5-celled capsules about 25 mm long. **Flowering** Spring and summer. **Habitat** Rainforest margins and swampy forest sites along the coast of NT, Qld and northeastern NSW. **Family** Malvaceae.

Alyogyne hakeifolia

Hibiscus splendens

Hibiscus heterophyllus

Hibiscus diversifolius

Alyogyne huegelii

Hibiscus tiliaceus

Hibbertia aspera

Low, prostrate or climbing shrub to 50 cm high with hairy, trailing branches. **Leaves** Spathulate to obovate, notched or pointed, hairy with curved-back margins, 3-20 mm long and 2-8 mm wide. **Flowers** Yellow, open, 8-12 mm across with 5 overlapping rounded and notched petals and 4-16 stamens united on one side of the flower, solitary on terminal stalks 5-15 mm long. **Flowering** Mainly in summer. **Habitat** Sandy soils in heaths and open forests of the coast and adjacent ranges in Qld, NSW, Vic., SA and northeastern Tas. **Family** Dilleniaceae.

Hibbertia linearis

Variable erect or trailing shrub to 2 m high. **Leaves** Flat, linear to oblong or obovate, sometimes pointed, 6-30 mm long and 1-6 mm wide. **Flowers** Yellow, open, 15-25 mm across with 5 overlapping rounded and notched petals and 15-25 free stamens. They are solitary, terminal or axillary and stalkless or with very short stalks. **Flowering** Spring and summer. **Habitat** Sandy soils in heaths and dry sclerophyll forests of the coast and adjacent ranges in southeastern Qld and NSW. **Family** Dilleniaceae.

Hibbertia serpyllifolia

Low or prostrate shrub with branches to 30 cm long. **Leaves** Linear to spathulate, sometimes notched, rough, with curved-back margins, 2-10 mm long and 1-3 mm wide. **Flowers** Yellow, open, 10-12 mm across with 5 overlapping rounded and notched petals and 15-20 free stamens, terminal and stalkless or with very short stalks. **Flowering** Spring and summer. **Habitat** Widespread in sandy soils in heaths of the coast and adjacent ranges in NSW, southeastern Vic. and northeastern Tas. **Family** Dilleniaceae.

Hibbertia fasciculata   Bundled Guinea Flower

Erect or straggling shrub to 40 cm high with softly-hairy young branches. **Leaves** Crowded and clustered, narrow-linear, stalkless, concave, slightly hairy, 4-6 mm long and about 1 mm wide. **Flowers** Yellow, open, 1-2 cm across with 5 overlapping rounded and sometimes notched petals and 8-12 stamens, terminal and stalkless on short axillary branches. **Flowering** From winter to early summer. **Habitat** Widespread on sandy soils in heaths and dry sclerophyll forests of the coast and ranges in southeastern Qld and NSW. **Family** Dilleniaceae.

Hibbertia riparia   Erect Guinea Flower

Low bushy shrub to 60 cm high, sometimes with dense bristly hairs on the stems. **Leaves** Alternate, narrow-linear with a prominent midrib and curled-under margins, hairy below, 4-12 mm long and 1-2 mm wide. **Flowers** Yellow, open, 12-24 mm across with 5 overlapping rounded and notched petals and 6-16 stamens, axillary or terminal on short stalks. **Flowering** Spring and summer. **Habitat** Widespread in sandy heaths and open forests in Qld, NSW, Vic., SA and Tas. **Family** Dilleniaceae.

Hibbertia procumbens   Spreading Guinea Flower

Low or prostrate shrub with spreading stems to 30 cm long. **Leaves** Narrow-linear to narrow-lanceolate, 15-20 mm long and about 2 mm wide. **Flowers** Yellow, open, 2-3 cm across with 5 overlapping, rounded to notched petals and about 20 curly yellow stamens, solitary, terminal and stalkless. **Flowering** Summer. **Habitat** Heaths in sandy soils on the central coast of NSW, and the coast and ranges of Vic. and Tas. **Family** Dilleniaceae.

Hibbertia linearis

Hibbertia riparia

Hibbertia aspera

Hibbertia fasciculata

Hibbertia procumbens

Hibbertia serpyllifolia

Hibbertia scandens
Climbing Guinea Flower

Climbing shrub with stems to 4 m long. **Leaves** Elliptic to lanceolate or obovate, usually stalkless and stem-clasping, silky-hairy below, 3-9 cm long and 1-3 cm wide. **Flowers** Yellow, open, 2-6 cm across with 5 overlapping rounded petals and more than 30 stamens. They are solitary in the leaf axils on stalks 2-4 mm long.
Flowering Most of the year. **Habitat** Widespread on coastal dunes, heaths, open forests and rainforests of the coast and tablelands in Qld, NSW and the northern NT. **Family** Dilleniaceae.

Goodenia glabra
Smooth Goodenia. Shiny Pansy

Prostrate annual or perennial herb with upright stems to 40 cm long, sometimes covered with cottony hairs.
Leaves Narrow-ovate to oblong, stalkless, lobed on one side near the base, sometimes toothed, 3-9 cm long and 7-11 mm wide. **Flowers** Yellow, sometimes with purplish markings, sometimes cottony inside, open, 10-18 mm long and about 3 cm across with 5 unequal overlapping lobes, arising from a short, slit tube, 2 lobes arch over the pollen cup. They are solitary or in 3-flowered leafy racemes on stalks to 35 mm long. **Fruits** Globular to ovoid capsules, 7-10 mm long. **Flowering** Year round. **Habitat** Widespread in dry open communities on sandy soils of the inland slopes and plains of Qld, NSW, SA, and southern NT. **Family** Goodeniaceae.

Petalostylis labicheoides
Butterfly Bush

Erect, rounded shrub to 3 m high with downy branches. **Leaves** Opposite, divided into 5-21 lanceolate to elliptic leaflets each 1-3 cm long and 3-8 mm wide. **Flowers** Yellow with red markings, open, 3-4 cm across with 5 overlapping rounded lobes and a protruding, petal-like, curved style, arranged in short racemes of 1-5 flowers.
Fruits Flat, oblong pods, 2-3 cm long with flat, shiny seeds. **Flowering** Mainly in spring. **Habitat** Sand plains, dune fields and rocky ridges inland in Qld, NSW, SA and WA. **Family** Caesalpiniaceae.

Sida petrophila *(syn. S. calyxhymenia)*
Rock Sida

Erect, often spindly, hairy shrub to 2 m high. **Leaves** Alternate, oblong to linear, 15-60 mm long and 4-30 mm wide with slightly toothed margins, covered with dense, greyish, small hairs. **Flowers** Yellow, open, about 2 cm across with 5 overlapping rounded lobes and a protruding column of yellow stamens, borne singly or in axillary or terminal clusters of 2-6 flowers on stalks 7-20 mm long. **Fruits** Dry, hairy and conical capsules, 5-7 mm across with 5-7 segments. **Flowering** Most of the year. **Habitat** Dry, rocky hillsides, usually in mulga communities inland in Qld, NSW, SA, WA and NT. **Family** Malvaceae.

Senna odorata *(syn. Cassia odorata)*

Sprawling shrub to 2.5 m high. **Leaves** Opposite, 5-15 cm long, divided into 12-26 lanceolate to elliptic leaflets 1-3 cm long and 2-10 mm wide with curved-back margins, paler below, with glands between each pair.
Flowers Yellow, open to cup-shaped, 20-25 mm across with 5 overlapping rounded lobes, arranged in loose axillary clusters of 2-6 flowers. **Fruits** Cylindrical pods, 7-12 cm long and 5-6 mm across. **Flowering** Spring and summer. **Habitat** Widespread in moist sclerophyll forests and rainforest margins along the coast and tablelands in Qld and NSW. **Family** Caesalpiniaceae.

Senna floribunda *(syn. Cassia floribunda)*
Smooth Cassia

Erect shrub to 3 m high. **Leaves** Opposite, 5-10 cm long, divided into 6-10 opposite, ovate leaflets, 3-7 cm long and 15-30 mm wide, with a narrow gland between the lowest pairs. **Flowers** Yellow, open to cup-shaped, about 4 cm across with 5 overlapping rounded lobes and 7 stamens, arranged in loose axillary or terminal racemes of 5-10 flowers on stalks to 4 cm long. **Fruits** Brown, cylindrical pods, 5-8 cm long and 7-12 mm across. **Flowering** From spring to autumn. **Habitat** Introduced from Mexico, naturalised and widespread in forests, moist gullies and rainforest margins of the coast and tablelands in southeastern Qld and NSW. **Family** Caesalpiniaceae.

Senna floribunda

Sida petrophila

Petalostylis labicheoides

Senna odorata

Hibbertia scandens

Goodenia glabra

Gossypium sturtianum
Sturt's Desert Rose

Floral emblem of the Northern Territory, a dense shrub to 2 m high with black spots on the stems. **Leaves** Alternate, orbicular to ovate or obovate, 25-60 mm long and 1-6 cm wide, dotted with tiny oil glands. **Flowers** Pink or lilac with a dark-red centre, open, 5-10 cm across with 5 overlapping rounded lobes and a protruding column of fused stamens, solitary with stalks 5-20 mm long. **Fruits** Black-spotted ovoid capsules, 20-25 mm long. **Flowering** Most of the year. **Habitat** Widespread in arid, rocky and sandy sites in gorges and gullies inland in all states. **Family** Malvaceae.

Lavatera plebeia
Native Hollyhock

Upright perennial herb to 4 m high, covered with small hairs. **Leaves** Alternate, broad-ovate to orbicular, 1-20 cm across, divided into 5-7 lobes with toothed margins, hairy on both surfaces, on hairy stalks 10-15 mm long. **Flowers** Lilac to pink or white, open, 3-6 cm across with 5 overlapping rounded lobes and a protruding column of fused stamens, axillary in clusters of 2-5 flowers, to 25 mm long. **Fruits** Dry capsules, 6-7 mm diameter, splitting into 8-16 segments. **Flowering** Most of the year. **Habitat** Sandy soils near watercourses, inland in all states. **Family** Malvaceae.

Pavonia hastata
Pink Pavonia

Low, spreading, hairy shrub to 1.5 m high. **Leaves** Alternate, oblong to lanceolate or heart-shaped, 1-6 cm long and 1-3 cm wide with toothed margins, rough above and hairy below, on stalks to 3 cm long. **Flowers** Pink to reddish purple with a deep red centre, open, about 3 cm across with 5 overlapping, rounded lobes and a protruding column of fused stamens, solitary on terminal or axillary stalks. **Fruits** Hairy, about 8 mm diameter. **Flowering** Spring and summer. **Habitat** Introduced from South America, naturalised and found in open forests and clearings of the coast and ranges of Qld, NSW, SA and northern WA. **Family** Malvaceae.

Melastoma affine (syn. M. polyanthum)

Low bristly shrub to 2 m high. **Leaves** Opposite, ovate to oblong or lanceolate, covered with stiff hairs, 4-12 cm long and 2-4 cm wide, with 5 prominent longitudinal veins. **Flowers** Pale-purple to white, open, 2-4 cm across with 5 overlapping, rounded to notched lobes, 5 long, curved stamens and 5 shorter stamens, arranged in terminal clusters of 5-11 flowers. **Fruits** Semi-succulent berries about 1 cm across. **Flowering** Spring and summer. **Habitat** Moist or swampy forests and rainforest margins along the coast of Qld and northern NSW. **Family** Melastomataceae.

Abutilon halophilum

Low, downy shrub to 50 cm high. **Leaves** Alternate, orbicular to heart-shaped with notched margins, 1-3 cm long and 1-2 cm wide, covered with soft hairs. **Flowers** Yellow, open to cup-shaped, 15-25 mm across with 5 overlapping, rounded lobes and a protruding column of fused yellow stamens, solitary on long axillary stalks. **Fruits** Hairy capsules 10-15 mm long. **Flowering** Winter, spring and summer. **Habitat** Red gravelly or sandy soils, inland in floodplains and saltbush communities of the NT, Qld, NSW and SA. **Family** Malvaceae.

Rubus hillii
Queensland or Molucca Bramble

Scrambling or climbing shrub with long, woolly, prickly stems. **Leaves** Alternate with 3-7 broad lobes and toothed margins, ovate to orbicular or heart-shaped, covered with white or rusty hairs below, often with prickles on the principal veins, 2-15 cm long and 3-10 cm wide on stalks 2-6 cm long. **Flowers** White or red, silky hairy, open, about 1 cm across with 5 overlapping, rounded lobes and 5 longer, pointed sepals and numerous stamens, arranged in irregular axillary clusters. **Fruits** Red, globular drupes about 12 mm diameter. **Flowering** Spring and summer. **Habitat** Widespread near streams in wet sclerophyll forests and rainforests of southeastern Qld, eastern NSW, southeastern Vic., southwestern WA and northeastern Tas. **Family** Rosaceae.

*Abutilon
halophilum*

*Lavatera
plebeia*

*Melastoma
affine*

*Pavonia
hastata*

Rubus hillii

Gossypium sturtianum

Howittia trilocularis   **Blue Howittia**

Erect, sometimes scrambling shrub to 3 m high with rusty hairs on the slender branches. **Leaves** Alternate, ovate to lanceolate, 2-10 cm long and 1-5 cm wide with toothed margins sometimes curved back, dark-green with scattered hairs above, with dense white or rusty hairs below. **Flowers** Violet to lavender, to 2 cm across with 5 overlapping rounded lobes and a protruding column of stamens, axillary in clusters of 1-3 flowers on long stalks. **Fruits** 3-valved capsules about 8 mm diameter. **Flowering** Winter, spring and early summer. **Habitat** Widespread on the coast and tablelands often in damp forest areas on sandy soils in NSW and Vic. **Family** Malvaceae.

Thysanotus patersonii   **Twining Fringe Lily**

Slender, twining or trailing perennial herb with many stems. **Leaves** Thread-like, 10-20 cm long, one or 2 produced annually, arising from the base of the plant and withering early, when the leafless quadrangular climbing stems become photosynthetic. **Flowers** Violet, open, 1-2 cm across with 6 overlapping lobes, 3 fringed and rounded, 3 narrow, they are solitary and terminal on one or 2 multi-branched, flowering stems, 10-100 cm long, produced annually. **Fruits** Globular capsules 3-4 mm across. **Flowering** Spring. **Habitat** Widespread, mainly in inland districts of all states. **Family** Liliaceae.

Thysanotus juncifolius   **Rush Fringe Lily**

Upright perennial herb to 65 cm high. **Leaves** Narrow-linear, 6-25 cm long, arising from the base of the plant, with up to 3 leaves produced annually. **Flowers** Blue to purple, open, 2-3 cm across with 3 broad fringed lobes and 3 narrow pointed lobes, usually in clusters of 1-5. Flowers at the ends of branched flowering stems 15-65 cm long. **Fruits** Cylindrical capsules about 5 mm long and 3 mm wide. **Flowering** From spring to autumn. **Habitat** Widespread on sandy soils in damp coastal heaths and sclerophyll forests of southeastern Qld, central eastern and southeastern NSW, Vic. and SA. **Family** Liliaceae.

Thysanotus tuberosus   **Common Fringe Lily**

Upright perennial herb to 80 cm high. **Leaves** Narrow-linear, 10-60 cm long, arising annually from the base of the plant. **Flowers** Blue to purple, open, 15-35 mm across with 3 broad fringed lobes and 3 narrow pointed lobes, arranged terminal clusters of 1-8. Flowers on long, branching stems, 20-80 cm long. **Fruits** Cylindrical 3-valved capsules 3-7 mm across. **Flowering** From spring to autumn. **Habitat** Widespread in dry sclerophyll forests, woodlands and heaths in Qld, NSW, Vic and southeastern SA. **Family** Liliaceae.

Eustrephus latifolius   **Wombat Berry. Orange Vine**

Climbing shrub with slender, wiry stems to 6 m long. **Leaves** Alternate, ovate to lanceolate with prominent parallel veins, 3-10 cm long and 3-35 mm wide. **Flowers** White to pale-pink or mauve, open, about 15 mm across with 6 spreading lobes in two whorls, the inner 3 fringed, and 6 stamens fused into a tube. They are arranged in clusters of 2-10 flowers in the upper leaf axils. **Fruits** Globular yellow-orange fleshy capsules, 1-2 cm diameter with numerous black seeds. **Flowering** Spring and summer. **Habitat** Widespread in moist places in sclerophyll forests, woodlands, heaths and rainforest margins of the coast, ranges and inland slopes of Qld, NSW and southeastern Vic. **Family** Philesiaceae.

Arthropodium milleflorum   **Pale Vanilla Lily**

Upright, tufted, perennial herb to 1.2 m high. **Leaves** arise from the base of the plant, grass-like, sometimes greyish, 3-60 cm long and 1-2 cm wide. **Flowers** Pale lilac to white or pink with a vanilla scent, open, about 2 cm across with 6 spreading, overlapping lobes, 3 larger with wavy margins, and 6 conspicuously bearded stamens, arranged in small clusters on slender stalks to 24-120 cm long. **Fruits** Dry, round capsules, 4-5 mm across. **Flowering** Mainly in spring. **Habitat** Widespread in hilly and rocky sites in southeastern Qld, NSW, Vic., southeastern SA and Tas. **Family** Liliaceae.

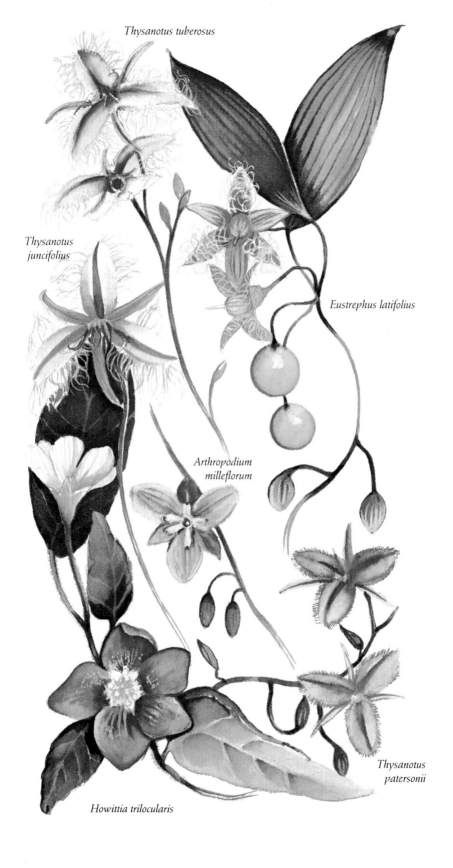

Thysanotus tuberosus

Thysanotus juncifolius

Eustrephus latifolius

Arthropodium milleflorum

Thysanotus patersonii

Howittia trilocularis

Dianella longifolia *(syn. D. laevis)* — Smooth or Pale Flax Lily

Upright tufted perennial herb to 1.5 m high. **Leaves** Long, linear, arising from and sheathing the base of the plant, 15-80 cm long and 2-25 mm wide. **Flowers** Pale blue to whitish, open, 1-2 cm across with 6 pointed lobes and 6 long, thick, pale-yellow anthers, arranged in a terminal panicle 30-90 cm long. **Fruits** Blue to purple berries 3-8 mm long. **Flowering** Spring and summer. **Habitat** Widespread in sclerophyll forests and woodlands on sandy soils from the coast to the western plains in all states. **Family** Liliaceae.

Dianella revoluta — Spreading Flax Lily

Tufted or mat-forming perennial herb to 1.2 m high. **Leaves** Long, linear, arising from and sheathing the base of the plant, leathery with curled-under margins, 10-85 cm long and 4-15 mm wide. **Flowers** Purple to blue, open, 7-12 mm across with 6 rounded lobes and 6 long, thick, yellow-brown anthers. They are arranged in a terminal panicle of 2-9 flowers, 30-90 cm long. **Fruits** Blue to purple berries 4-10 mm long. **Flowering** Winter, spring and summer. **Habitat** Widespread in sclerophyll forests, woodlands and mallee in all states. **Family** Liliaceae.

Stypandra glauca — Nodding Blue Lily

Upright, tufted, perennial herb to 1.5 m high and 1 m across at the base, often flowering below 30 cm tall. **Leaves** Linear, grass-like, sheathing the base of the plant, alternate, grey-green, 5-20 cm long and 1-15 mm wide. **Flowers** Deep blue to white, open, 2-3 cm across with 6 pointed lobes and 6 conspicuous yellow, hairy, protruding stamens. They are arranged in loose racemes on slender, bent stalks. **Fruits** Ovoid capsules with 3 ridges, 3-12 mm long with black seeds. **Flowering** Winter and spring. **Habitat** Widespread in sclerophyll forests and woodlands on poorer sandy soils in southeastern Qld, the coast and ranges of NSW, Vic., central southern SA and central and southwestern WA. **Family** Liliaceae.

Schelhammera undulata — Lilac Lily

Erect or scrambling herb to 20 cm high. **Leaves** Broad-lanceolate to ovate with wavy margins, 2-5 cm long and 7-18 mm wide, with short stalks or stalkless. **Flowers** Pale lilac to pink, open, 10-20 mm across with 6 pointed lobes and 6 stamens, solitary or in small terminal or axillary clusters on slender stalks 1-3 cm long. **Fruits** Globular, wrinkled capsules 5-8 mm across. **Flowering** Spring. **Habitat** Widespread in shady sites in open forests of coastal NSW and southeastern Vic. **Family** Liliaceae.

Tripladenia multiflora *(syn. Kreyssigia multiflora)* — Sarsparilla Lily

Spreading perennial herb to 40 cm high with many slender stems. **Leaves** Stem-clasping, lanceolate to ovate or heart-shaped, 4-9 cm long and 10-45 mm wide, shiny dark-green above with prominent veins. **Flowers** Pink to pale purple, open, 15-25 mm across with 6 oblong lobes, solitary or in small axillary clusters on slender stalks 3-6 cm long. **Fruits** Ovoid, wrinkled capsules, 6-8 mm across with yellow or brown seeds. **Flowering** Spring and summer. **Habitat** Locally abundant in rainforests and wet sclerophyll forests of southeastern Qld and northeastern NSW. **Family** Liliaceae.

Sowerbaea juncea — Vanilla Plant. Rush Lily

Upright, rush-like, perennial herb to 75 cm high. **Leaves** Blue-green, grass-like to cylindrical, 5-50 cm long and 1-2 mm wide, arising from the base of the plant. **Flowers** Pink-lilac to white with a vanilla perfume, open, 15-35 mm across with 6 pointed lobes and protruding yellow stamens. They are arranged in terminal clusters of up to about 20 flowers. **Fruits** 3-celled capsules 2-3 mm long. **Flowering** Year round. **Habitat** Widespread in waterlogged sites in coastal heaths and tablelands in southeastern Qld, NSW, southeastern Vic. and northeastern Tas. **Family** Liliaceae.

Stypandra glauca

*Dianella
longifolia*

*Dianella
revoluta*

*Schelhammera
undulata*

*Sowerbaea
juncea*

Tripladenia multiflora

Thelionema caespitosum
Blue or Tufted Lily
(syn. *Stypandra caespitosa*)

Upright tufted perennial herb to 90 cm high with thin, branching stems. **Leaves** Grass-like, 10-45 cm long and 2-12 mm wide, arising from the base of the plant. **Flowers** Blue or cream, star-like, 16-26 mm across with 6 lobes and 6 protruding yellow stamens, arranged in loose terminal clusters. **Fruits** Oblong capsules 4-10 mm long. **Flowering** Spring and summer. **Habitat** Widespread on wet, sandy soils in coastal heaths and nearby ranges of southeastern Qld, NSW, Vic. and southeastern SA. **Family** Liliaceae.

Orthrosanthus multiflorus
Morning Flag

Upright perennial herb to 60 cm high. **Leaves** Linear, arising from and sheathing the base of the plant, 16-50 cm long and 2-6 mm wide. **Flowers** Blue to purple, 2-4 cm across with 6 rounded spreading lobes and 3 stamens, arranged in leafy racemes of 3-7 flowers. **Fruits** Capsules 12-20 mm long, containing globular seeds about 2 mm diameter. **Flowering** Spring. **Habitat** Coastal heaths and scrub on the south coast of SA, southwestern Vic and southern WA. **Family** Iridaceae.

Dichopogon strictus (syn. *Arthropodium strictum*)
Chocolate Lily

Upright perennial herb, chocolate scented, to 1 m high. **Leaves** Linear, grass-like, 7-65 cm long and 1-12 mm wide, arising from the base of the plant. **Flowers** Blue-violet, rarely white with purple-tipped stamens, open, 15-28 mm across with 3 broad, delicately crinkled lobes and 3 narrower lobes, arranged in racemes 20-100 cm long on slender, drooping branches. **Fruits** Globular capsules 4-8 mm across. **Flowering** From winter to summer. **Habitat** Widespread in grasslands and open forests in Qld, NSW, Vic., and southwestern WA. **Family** Liliaceae.

Caesia parviflora (syn. *C. vittata*)
Pale Grass Lily

Upright tufted perennial herb to 60 cm high. **Leaves** Linear, grass-like, to 40 cm long and 1-8 mm wide, arising from the base of the plant and sometimes withering early. **Flowers** Blue to lilac, rarely greenish-white, open, to 6-18 mm across with 6 spreading, pointed lobes and long, yellow-tipped stamens, arranged in clusters of 2-6 flowers along the upper stem. **Fruits** 3-lobed capsules 2-5 mm wide. **Flowering** Spring and summer. **Habitat** Widespread in heaths, dry sclerophyll forests and woodlands in all states except WA and the NT. **Family** Liliaceae.

Bauera rubioides
Wiry Bauera

Scrambling, prostrate or bushy shrub to 2 m high with wiry, hairy branches to 3 m long. **Leaves** Opposite, stalkless, divided into 3 usually toothed leaflets, 4-15 mm long and 2-3 mm wide, oblong to lanceolate, dark-green above, paler below with scattered hairs. **Flowers** Pink to white, 13-20 mm across with 6-8 spreading rounded petals and 30-90 yellow stamens, solitary and axillary on hairy stalks to 2 cm long. **Fruits** 2-valved capsules. **Flowering** Most of the year. **Habitat** Widespread in damp areas in heaths and forests of the coast and ranges in southeastern Qld, central and southern NSW, Vic., southeastern SA and Tas. **Family** Saxifragaceae.

Bauera sessiliflora
Grampians Bauera

Scrambling shrub with wiry, hairy branches to 2 m long. **Leaves** Opposite, stalkless, divided into 3 hairy leaflets, 6-25 mm long, oblong to lanceolate, dark-green above, paler below. **Flowers** Pink to magenta, open, 10-18 mm across usually with 6 spreading rounded lobes and numerous black-tipped stamens, stalkless, in small axillary clusters. **Fruits** 2-valved capsules. **Flowering** Spring and summer. **Habitat** Sheltered sandy hollows near streams in the Grampians, Vic. **Family** Saxifragaceae.

*Orthrosanthus
multiflorus*

Bauera rubioides

*Bauera
sessiliflora*

Dichopogon strictus

Caesia parviflora

Thelionema caespitosum

M.westmacott.

Libertia paniculata
Branching Grass Flag
Upright, densely tufted, perennial herb to 60 cm high. **Leaves** Long, linear, leathery, arising from and sheathing the base of the plant, 20-60 cm long and 4-12 mm wide. **Flowers** White, with stalks to 1 cm long, open, about 2 cm across with 6 delicate lobes, 3 larger than the others, and 3 long yellow-tipped stamens. they are arranged in long, loose, oblong panicles of 3-6 flowers. **Fruits** Ovoid, black, 3-valved capsules, 5-8 mm across, with dark brown, angular seeds, about 1 mm across. **Flowering** Spring. **Habitat** Widespread in rainforests and damp gullies in wet sclerophyll forests of coast and ranges of southeastern Qld, NSW and southeastern Vic. **Family** Iridaceae.

Bulbine semibarbata
Wild Onion. Leek Lily
Upright, annual or sometimes perennial herb to 60 cm high. **Leaves** Grass-like, linear, channelled, succulent, arising from the base of the plant, 5-30 cm long and up to 5 mm wide. **Flowers** Yellow, open, 1-2 cm across with 6 pointed lobes and 6 long, red to yellow stamens, some or all of which are bearded, arranged in a raceme to 2-23 cm long on stalks 13-20 cm long. **Fruits** Capsules 2-8 mm long. **Flowering** From winter to summer. **Habitat** Widespread around salt lakes and granite outcrops from the coast to the interior in all states except the NT. **Family** Liliaceae.

Bulbine bulbosa
Bulbine Lily. Golden Lily. Native Leek
Upright perennial herb to 75 cm high. **Leaves** Broad linear, channelled, succulent, grass-like, arising from the base of the plant, 5-50 cm long and up to 8 mm wide. **Flowers** Yellow, open, often perfumed, 2-4 cm across with 6 pointed lobes and 6 yellow stamens, arranged in racemes of up to 50 flowers, 5-26 cm long, on stalks 19-50 cm long, with numerous unopened buds at the tip. **Fruits** Capsules 3-6 mm long with dark brown seeds to 2 mm long. **Flowering** Spring and summer. **Habitat** Widespread in damp sites in woodlands, sclerophyll forests and grasslands on rocky or sandy soils in all states except WA and the NT. **Family** Liliaceae.

Burchardia umbellata
Milkmaids
Erect, honey-scented, perennial herb to 60 cm high. **Leaves** Narrow-linear, concave, some sheathing the base of the plant, to 60 cm long and 4 mm wide. **Flowers** White to pale-pink, open, about 10-16 mm across with 6 pointed lobes and 6 purple-tipped stamens surrounding a prominent red ovary, arranged in terminal clusters of 2-9 flowers on stalks 10-65 cm high. **Fruits** Ovoid 3-angled capsules 10-15 mm long. **Flowering** Spring. **Habitat** Sandy soils in swamps, heaths and open forests of southeastern Qld, the coast and tablelands of NSW, Vic., SA, southwestern WA and Tas. **Family** Liliaceae.

Wurmbea dioica (syn. Anguillaria dioica)
Early Nancy
Upright perennial herb, sweetly perfumed, to 30 cm high. **Leaves** Grass-like, linear to lanceolate, 15-35 cm long and up to 5 mm wide, stem-sheathing at the base. **Flowers** White, pink or greenish-white with faint purple borders, 10-25 mm across with 6 rounded lobes and 6 long, yellow or purple-tipped stamens, stalkless in a terminal open spike of 1-11 flowers. **Fruits** Brown 3-celled angular capsules, 5-10 mm long. **Flowering** Spring. **Habitat** Widespread on well-drained soils in forests and woodlands, inland on sand dunes and grasslands in all states except WA and the NT. **Family** Liliaceae.

Geitonoplesium cymosum
Scrambling Lily
Climbing shrub with tangled, wiry branches to 8 m long. **Leaves** Alternate, linear to narrow-ovate to lanceolate with prominent parallel veins, shiny-green above, 2-10 cm long and 3-30 mm wide. **Flowers** White inside and purple-green outside, open, about 2 cm across with 6 spreading lobes and 6 long, yellow-tipped stamens, arranged in drooping terminal clusters. **Fruits** Dark blue to black globular berries 6-20 mm diameter with numerous black seeds. **Flowering** Spring and summer. **Habitat** Widespread in moist areas in or near rainforests, sclerophyll forests and woodlands of the coast and ranges of Qld, NSW and southeastern Vic. **Family** Philesiaceae.

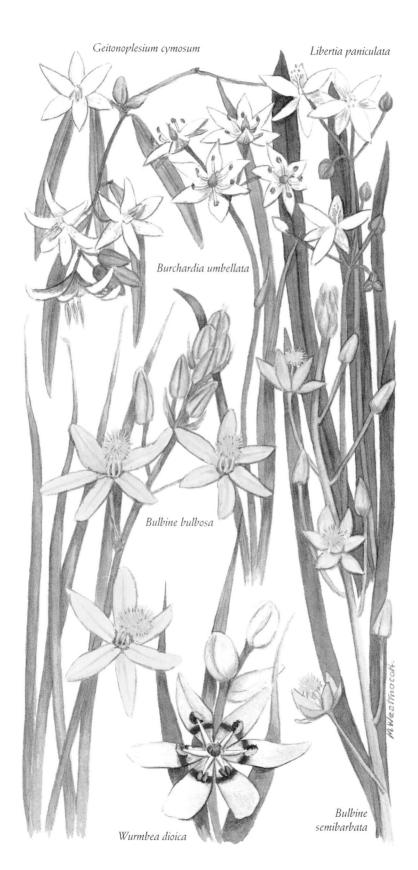

Geitonoplesium cymosum

Libertia paniculata

Burchardia umbellata

Bulbine bulbosa

M. Westmacott

Wurmbea dioica

Bulbine semibarbata

Nymphaea gigantea
Giant or Blue Waterlily

Aquatic perennial herb with large floating leaves and upright flowering stem. **Leaves** Heart-shaped to ovate or orbicular, 10-75 cm across, with regularly toothed margins, purplish below with prominent veins. **Flowers** Lotus-like, blue to mauve, pink or white, solitary, open, 6-30 cm across with many petals in whorls around numerous yellow stamens, on long stems standing to 50 cm above the water. **Fruits** Spongy berries with red, turning grey seeds. **Flowering** Most of the year. **Habitat** Still water in rivers and ponds along the east coast from northern NSW to northern Qld, NT and northern WA. **Family** Nymphaeaceae.

Xanthosia rotundifolia
Southern Cross

Upright perennial herb to 60 cm high. **Leaves** Alternate, orbicular to ovate or wedge-shaped with toothed margins, leathery, 3-5 cm long. **Flowers** White to yellow inside, purplish outside, open, arranged in 4-flowered terminal clusters resembling flattened crosses, 3-6 cm across, consisting of 3-lobed flowers with long petal-like bracts below the 4 stalks. **Flowering** Winter and spring. **Habitat** Sandy and rocky sites of woodlands and heaths in southwestern WA. **Family** Apiaceae.

Passiflora herbertiana
Native Passion Flower

Climbing shrub with long stems. **Leaves** Alternate, 3-lobed, ovate to heart-shaped, covered with small hairs, sometimes with toothed margins, 4-12 cm long and 4-11 cm wide. **Flowers** Pale orange to green, open, about 6 cm across with 5 spreading, narrow lobes, 5 stamens united with the long protruding stalk of the ovary, and 3 long styles. They are solitary in the leaf axils on stalks 20-35 mm long. **Fruits** Green, rather dry, ellipsoid berries, 4-7 cm long. **Flowering** Spring and summer. **Habitat** Fertile soils in forests of the coast and ranges in Qld and NSW. **Family** Passifloraceae.

Nelumbo nucifera
Sacred Lotus. Lotus Lily

Aquatic perennial herb anchored in water 1-2 m deep. **Leaves** Orbicular, 20-100 cm across on long, thick stalks usually well above the surface. **Flowers** Red to pink, white or yellow, open, 15-30 cm across with 12-27 large lobes and numerous yellow stamens, solitary on long stalks. **Fruits** Hard ellipsoid achenes about 2 cm long. **Flowering** Winter. **Habitat** Lagoons in coastal areas of northern WA, northern NT and Qld. **Family** Nelumbonaceae.

Ranunculus anemoneus
Snow or Anemone Buttercup

Upright perennial herb to 60 cm high. **Leaves** Stalkless, alternate or arising from the base of the plant, leathery, orbicular, 2.5-11 cm long and 4-13 cm wide, deeply divided into 3-5 linear to lanceolate lobed segments. **Flowers** White with pink buds, open, 2.5-6 cm across with 15-35 overlapping, spreading petals and numerous yellow stamens, terminal or axillary on 1-3 flowered leaf-opposing stalks to 13 cm long. **Fruits** Smooth achenes with or without a straight beak. **Flowering** Spring and early summer. **Habitat** Steep, rocky, alpine herbfields and subalpine woodlands subject to snowfalls in the southern tablelands of NSW. **Family** Ranunculaceae.

Proiphys cunninghamii (syn. Eurycles cunninghamii)

Upright, bulbous, perennial herb to 60 cm high. **Leaves** Ovate to heart-shaped, arising from the base of the plant, 10-25 cm long and 6-13 cm wide on long fleshy stems. **Flowers** White, shortly tubular, 10-20 mm long and 3-4 cm across with 6 overlapping, spreading lobes and 6 stamens, arranged in clusters of 5-12 flowers at the end of a long fleshy stem. **Fruits** Globular, orange-red and fleshy, 1-3 cm across. **Flowering** Spring and early summer. **Habitat** Rocky hills in wet sclerophyll forests and rainforest margins of southeastern Qld and northeastern NSW. **Family** Amaryllidaceae/Liliaceae.

Nelumbo nucifera

Proiphys cunninghamii

Nymphaea
gigantea

Passiflora
herbertiana

Xanthosia rotundifolia

Ranunculus
anemoneus

Carpobrotus glaucescens   Angular Pig Face. Coastal Noonflower

Prostrate perennial herb with stems to 2 m long. **Leaves** Thick, fleshy, triangular in section, blue-green, 3-10 cm long and 9-15 mm wide with a small pointed tip. **Flowers** Purple with an orange-yellow centre, almost stalkless, 3-6 cm across with 100-150 lobes in 3-4 rows, with white bases spreading out from a disc with 300-400 stamens, solitary and terminal. **Fruits** Red to purplish cylindrical berries, 2-3 cm long. **Flowering** Most of the year. **Habitat** Coastal sand dunes of southern Qld, NSW and southeastern Vic. **Family** Aizoaceae.

Carpobrotus aequilaterus

Prostrate perennial herb with stems to 2 m long. **Leaves** Thick, fleshy, triangular in section, dull-green or blue-green, 3-9 cm long and 5-12 mm wide. **Flowers** Purple with a yellow to pink centre, open, 25-80 mm across with 70-150 lobes in 3 rows, spreading out from a disc with 150-400 stamens, solitary and terminal. **Fruits** Purple ovoid berries, 15-30 mm long. **Flowering** Spring and summer. **Habitat** Introduced from South America, established in sandy coastal areas of southeastern Qld, NSW, Vic. SA, southwestern WA, western and southwestern Tas. **Family** Aizoaceae.

Stellaria flaccida

Soft, prostrate or ascending perennial herb with angular stems to 50 cm long, often with curly hairs. **Leaves** Opposite, ovate to lanceolate, stalkless and rigid, sharply-pointed, 5-18 mm long and 2-10 mm wide with wavy margins. **Flowers** White or pink, 1-2 cm across with 10 spreading lobes and 10 long, red-tipped stamens, solitary in the leaf axils on a stalk 2-4 cm long. **Fruits** Dry, ovoid, 3-4 valved capsules. **Flowering** Spring and summer. **Habitat** Rainforest margins, moist gullies of the coast and tablelands in southeastern Qld, NSW, Vic. and northeastern Tas. **Family** Caryophyllaceae.

Disphyma crassifolium   Round-leaved Pig Face

Prostrate annual or short-lived perennial herb with stems to 1 m long. **Leaves** Opposite, succulent, triangular to cylindrical in section, club-shaped, 2-5 cm long and 4-10 mm wide, pale-green, often reddish at the tip. **Flowers** Pink to magenta with a white centre, 2-5 cm across with many narrow lobes surrounding a central disc, solitary and terminal on slender stalks, 2-10 cm long. **Fruits** Globular 5-valved capsules to 12 mm across. **Flowering** Spring and summer. **Habitat** Salt tolerant, abundant on coastal headlands and inland in mallee scrubs of southeastern Qld, the central and southern coast and western plains of NSW, Vic., SA, the southern half of WA and Tas. **Family** Aizoaceae.

Boronia parviflora   Swamp Boronia

Erect or low shrub with weak stems to 1 m long. **Leaves** Opposite, thick, narrow-elliptic or very narrow ovate, often finely-toothed near the tip, often reddish below, 7-25 mm long and 2-8 mm wide. **Flowers** Pale to deep pink, open, 5-10 mm across with 8 spreading lobes and 8 stamens, terminal on stalks 2-10 mm long in clusters of 1-3 flowers. **Fruits** Hairless capsules, exploding when ripe. **Flowering** Most of the year. **Habitat** Widespread in sandy and peaty soils, often at the edge of swamps of the coast and tablelands in southeastern Qld, NSW, Vic., southeastern SA and Tas. **Family** Rutaceae.

Actinotus helianthi   Flannel Flower

Upright annual or perennial herb to 1 m high, covered with woolly hair. **Leaves** Scattered and variable, to 10 cm long and 7 cm wide, divided into 2-3 linear to oblong segments, 15-30 mm long and about 5 mm wide, either entire or divided again into 2-3 lobes, grey green and hairy above, whitish hairy below. **Flowerheads** White to cream, sometimes tipped with green, comprising a globular cluster of many tiny florets surrounded by a ring of 10-18 white, petal-like bracts, open, 3-8 cm across, densely covered in soft hair. **Fruits** Ovate, hairy, compressed capsules 3-5 mm long and 2-3 mm wide. **Flowering** Year round. **Habitat** Widespread in open forests, heaths and sand dunes along the coast and ranges in southeastern Qld and NSW. **Family** Apiaceae.

Stellaria flaccida

Boronia parviflora

Actinotus helianthi

Carpobrotus aequilaterus

Disphyma crassifolium

Carpobrotus glaucescens

Thryptomene calycina
Grampians Thryptomene

Dense, spreading shrub to 3 m high with long slender branches. **Leaves** Opposite, elliptic, stalkless, leathery, 4-12 mm long and 2-3 mm wide, aromatic, dotted below with oil glands. **Flowers** White with red or yellow centres, open to cup-shaped, 3-6 mm across with 10 overlapping, rounded lobes. They are arranged in axillary pairs in profuse clusters towards the ends of branches. **Fruits** Cup-shaped capsules. **Flowering** Winter and spring. **Habitat** Abundant in sandy soils in the Grampians and nearby areas of western Vic. and southeastern SA. **Family** Myrtaceae.

Polygala myrtifolia
Myrtle-leaf Milkwort

Dense erect shrub to 2.5 m high. **Leaves** Alternate, spirally arranged, elliptic to obovate, pale-green, 1-5 cm long and 4-15 mm wide. **Flowers** Deep-pink to purple, pea-like, 25-30 mm across with 2 widely-spreading lobes and a curved keel 12-15 mm long with a white crest, arranged in short terminal racemes on stalks about 1 cm long. **Fruits** Compressed capsules, 7-10 mm diameter. **Flowering** Most of the year, but mainly in spring. **Habitat** Introduced from South Africa, established along the coast of NSW, Vic., SA, WA and Tas. **Family** Polygalaceae.

Comesperma defoliatum
Leafless Milkwort

Erect wiry shrub to 60 cm high. **Leaves** Very small elliptic, 5-12 mm long and about 1 mm wide, on the lower part of the plant. **Flowers** Blue, pea-like, 3-5 mm long with 2 rounded spreading lobes, a central keel and 2 small yellowish-white lobes attached to the tube formed by the fused stamens. They are arranged in terminal racemes up to 10 cm long. **Fruits** Wedge-shaped capsules, 7-11 mm long. **Flowering** Spring and summer. **Habitat** Wet heaths, swamps in sclerophyll forests, on sandy soils of the coast and tablelands in southeastern Qld, NSW, Vic. and northeastern Tas. **Family** Polygalaceae.

Comesperma ericinum
Heath Milkwort

Erect shrub to 1.5 m high with slender, hairy stems. **Leaves** Crowded, alternate, spirally arranged, usually with curled-under margins, linear to oblong elliptic, paler below with a distinct midvein, 5-25 mm long and 1-4 mm wide. **Flowers** Pink to magenta, pea-like, 4-8 mm long with 2 rounded spreading lobes and a central keel partially enclosing the stamens. They are arranged in short terminal racemes. **Fruits** Wedge-shaped capsules, 7-8 mm long. **Flowering** Mainly in spring and summer. **Habitat** Dry sclerophyll forests and woodlands, usually on sandstone, along the coast and ranges of southeastern Qld, NSW, Vic. and northeastern Tas. **Family** Polygalaceae.

Mirbelia rubiifolia

Erect diffuse shrub with slender angular branches to 1 m long. **Leaves** Opposite in whorls of 3 with curved-back margins, sharply-pointed, narrow-ovate to lanceolate or linear with conspicuous veins, 8-25 mm long and 1-4 mm wide. **Flowers** Pink to purple, rarely white, pea-shaped, 8-9 mm long, arranged in axillary clusters and short terminal racemes. **Fruits** 2-celled ovoid pods about 5 mm long. **Flowering** Spring and early summer. **Habitat** Widespread in sandy heaths and sclerophyll forests of the coast and tablelands in southern Qld, NSW and southeastern Vic. **Family** Fabaceae.

Indigofera australis
Austral Indigo

Erect, spreading shrub to 2.5 m tall with slender, stiff stems with flattened hairs. **Leaves** Alternate, 4-10 cm long, pinnately divided into 5-25 oval to oblong leaflets, 5-40 mm long and 2-9 mm wide. **Flowers** Pink to lilac with dark brown hairs on the calyx, pea-shaped, 6-10 mm long with a standard 6-9 mm long, arranged in axillary racemes 5-15 cm long. **Fruits** Brown cylindrical pods 20-45 mm long. **Flowering** Winter and spring. **Habitat** Widespread in eucalypt forests and woodlands in all states except the NT. **Family** Fabaceae.

Polygala myrtifolia

Comesperma ericinum

Indigofera australis

Comesperma defoliatum

Thryptomene calycina

Mirbelia rubiifolia

Hovea elliptica
Tree Hovea

Erect slender shrub to 4 m high with rusty-hairy branches. **Leaves** Alternate, elliptical to lanceolate, stiff, dark-green above sometimes with rusty hairs below, 3-8 cm long and 1-4 cm wide. **Flowers** Blue to purple, pea-shaped, 12-15 mm across with a nearly orbicular, notched, standard petal and rusty hairs outside the calyx, arranged in short racemes. **Fruits** Globular pods 8-12 mm across. **Flowering** Spring. **Habitat** Karri and jarrah forests in southwestern WA. **Family** Fabaceae.

Hovea longifolia
Long-leaf Hovea

Erect shrub to 3 m high, with stems and branches densely covered with soft, curled grey or rusty hairs. **Leaves** Alternate, linear to lanceolate or narrow-elliptic, 2-8 cm long and 2-12 mm wide, with curved-back margins, arched up from the prominent midrib, stiff, dark-green above and sparsely covered with rusty hairs below. **Flowers** Blue to mauve, pea-shaped, about 4 mm long with a nearly orbicular, notched, standard petal with a pale-yellow central patch, and dense rusty hairs outside the calyx. They are arranged in small clusters of 1-3 stalkless flowers, or short racemes of 5-6 flowers. **Fruits** Globular or ovoid pods 8-15 mm long with seeds about 3 mm long. **Flowering** Mainly in late winter and spring. **Habitat** Widespread in shady sites along creek slopes in dry sclerophyll forests of the coast and ranges in NSW, southeastern Vic., eastern SA and Tas. **Family** Fabaceae.

Hovea pungens
Devil's Pins

Erect shrub to 2 m high with spindly softly-hairy branches. **Leaves** Alternate, rigid, narrow-linear or lanceolate with curved-back margins, sharp points and a prominent midrib, stiff, 1-2 cm long. **Flowers** Blue to purple, pea-shaped, about 1 cm across with a nearly orbicular, notched, standard petal and dense rusty hairs outside the calyx, arranged in profuse axillary clusters. **Fruits** Globular pods 8-12 mm across. **Flowering** Spring. **Habitat** Granite outcrops, coastal limestone and sandy heaths in southwestern WA. **Family** Fabaceae.

Hovea rosmarinifolia
Mountain Beauty

Erect shrub to 3 m high with slender branches covered with grey-brown hairs. **Leaves** Alternate, linear to oblong-elliptical with curled-under margins, dark-green and rough above with a prominent midrib, pale-green, scattered with light brown hairs below, stiff, 1-3 cm long and 1-3 mm wide. **Flowers** Mauve to lavender, pea-shaped, 5-9 mm long, with a nearly orbicular, notched, standard petal and dense, rusty hairs outside the calyx, usually in axillary pairs forming profuse, stalkless clusters. **Fruits** Ovoid pods, 8-12 mm long, with dense rusty hairs. **Flowering** Spring. **Habitat** Scattered in open forests on poor sandy soils at moderate elevations in the slopes and tablelands of NSW and Vic. **Family** Fabaceae.

Hovea trisperma
Western Hovea

Spreading or trailing shrub to 60 cm high. **Leaves** Alternate, linear to lanceolate or ovate with slightly curved-back margins, stiff and pointed, 3-8 cm long and about 1 cm wide. **Flowers** Violet to deep blue, pea-shaped, about 1 cm across with a nearly orbicular, notched, standard petal and dense rusty hairs outside the calyx, growing in long leafy racemes. **Fruits** Globular or ovoid pods 8-12 mm long. **Flowering** Winter and spring. **Habitat** Sandy soils in heaths and woodlands in southwestern WA. **Family** Fabaceae.

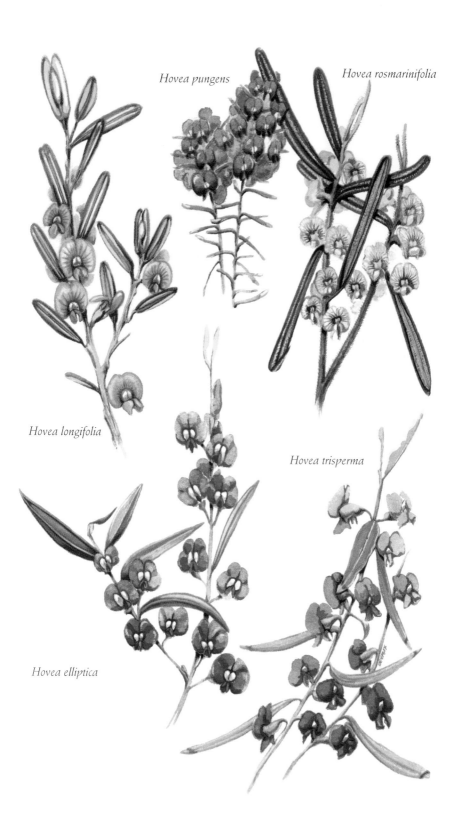

Hovea pungens

Hovea rosmarinifolia

Hovea longifolia

Hovea trisperma

Hovea elliptica

Hovea acutifolia

Erect shrub to 4 m high with dense grey or rusty hairs on the branches. **Leaves** Alternate, narrow-elliptic or lanceolate with slightly curved-back margins, fairly stiff, covered with rusty hairs below, 2-9 cm long and 3-20 mm wide. **Flowers** Mauve to blue, pea-shaped, about 1 cm long with a nearly orbicular, notched, standard petal and dense rusty hairs outside the calyx, growing in small axillary clusters of 1-4 flowers. **Fruits** Hairy globular or ovoid pods, 8-15 mm long. **Flowering** Winter and spring. **Habitat** Damp, sheltered sites and rainforest margins of the central and north coasts of NSW and eastern Qld. **Family** Fabaceae.

Hovea linearis
Common Hovea

Erect or decumbent slender shrub to 50 cm high with densely hairy young branches. **Leaves** Alternate with sparse hairs on the lower surfaces. Lower leaves are ovate to elliptic, 10-25 mm long and 5-10 mm wide. Upper leaves are linear to lanceolate, 25-60 mm long and 2-4 mm wide. **Flowers** Mauve to blue, pea-shaped, 6-10 mm long with a nearly orbicular, notched, standard petal and dense rusty hairs outside the calyx, growing in small axillary clusters of 1-2 flowers. **Fruits** Globular or ovoid pods, 7-9 mm long. **Flowering** Mainly in winter and spring. **Habitat** Widespread in eucalypt forests of the coast and tablelands and inland slopes in Qld, NSW, Vic., southeastern SA and Tas. **Family** Fabaceae.

Hardenbergia comptoniana
Native Wisteria. Wild Sarsparilla

Vigorous climbing shrub, reaching heights of 10 m. **Leaves** Alternate, divided into 3-5 narrow-ovate to lanceolate leaflets 5-12 cm long. **Flowers** Mauve to blue, pink or white, pea-shaped, about 9 mm long with a pair of white or greenish spots on the standard petal, arranged in long racemes. **Flowering** Winter and spring. **Habitat** Forests and bushland in southwestern WA. **Family** Fabaceae.

Hardenbergia violacea
False Sarsparilla. Native Lilac

Erect or climbing shrub with stems to 2 m long. **Leaves** Alternate, narrow-ovate to lanceolate, leathery and glossy-green above, paler below, 3-10 cm long and 1-6 cm wide. **Flowers** Violet with a pair of yellow spots on the notched standard petal, pea-shaped, about 1 cm long, arranged in terminal racemes, the upper ones often form a panicle with 20-30 flowers. **Fruits** Flat, oblong, brown pods with 6-8 seeds, 2-5 cm long and 8 mm wide. **Flowering** Mostly in spring. **Habitat** Widespread in a range of habitats, often in rocky sites in Qld, NSW, Vic., SA and northeastern Tas. **Family** Fabaceae.

Gompholobium scabrum
Painted Lady

Erect, slender shrub to 3 m high with many branches. **Leaves** Crowded in groups of 3, narrow-linear with margins curled-under to the midrib, to 15 mm long. **Flowers** Rose pink to mauve, pea-shaped, about 15 mm across, solitary and profuse on short stalks arising from the upper leaf axils. **Fruits** Globular pods with 2 seeds, about 5 mm diameter. **Flowering** Late winter and spring. **Habitat** Sandy soils in heaths and forests in southwestern WA. **Family** Fabaceae.

Psoralea patens
Native Verbine. Bullamon Lucerne

Prostrate or rarely erect hairy shrub with stems to 1 m long. **Leaves** Alternate, 3-75 mm long, divided into 3 ovate to lanceolate or oblong leaflets with toothed margins, sometimes with small hairs, 1-4 cm long and 5-25 mm wide. **Flowers** Pink to purple or blue, pea-shaped, 6-12 mm long with a silky-hairy calyx, on very short stalks, arranged in dense racemes to 27 cm long. **Fruits** Ovoid, hairy, blackish pods about 3 mm long. **Flowering** Mostly in winter and spring. **Habitat** Flood plains and watercourses on sandy soils inland in all mainland states. **Family** Fabaceae.

Gompholobium scabrum

Hovea acutifolia

Psoralea patens

Hardenbergia violacea

Hovea linearis

Hardenbergia comptoniana

Glycine clandestina
Twining Glycine

Twining or prostrate perennial herb with slender, hairy stems to 80 cm long. **Leaves** Alternate, hairy, divided into 3 linear to lanceolate leaflets, 1-8 cm long and 2-12 mm wide. Leaflets of lower leaves are broad obovate to circular, 5-30 mm long and 2-8 mm wide. **Flowers** Pink to mauve or white, pea-shaped, 6-10 mm long with a notched, almost orbicular standard petal, arranged in axillary racemes of 4-18 flowers. **Fruits** Oblong pods, 12-55 mm long and 2-4 mm wide with 4-12 seeds. **Flowering** Year round. **Habitat** Widespread in many situations, from the coast to the subalps and inland in all states except the NT. **Family** Fabaceae.

Swainsona galegifolia
Smooth Darling Pea

Erect or decumbent perennial shrubby herb to 1 m high with long stems, poisonous to stock. **Leaves** Alternate, 5-10 cm long, pinnately divided into 21-29 elliptic to narrow obovate leaflets 5-20 mm long and 2-8 mm wide. **Flowers** Pink to dark red, white, orange, yellow or purple, pea-shaped, 12-16 mm long with a notched orbicular standard petal, arranged in axillary racemes of 15-20 flowers. **Fruits** Elliptic pods, 15-50 mm long. **Flowering** Spring and early summer. **Habitat** Widespread in woodlands and grasslands of the coast, ranges and inland plains in Qld, NSW and northeastern Vic. **Family** Fabaceae.

Swainsona greyana
Hairy Darling Pea

Upright perennial shrubby herb to 1.5 m high with woolly stems, poisonous to stock. **Leaves** Alternate, 10-15 cm long, pinnately divided into 17-21 narrow-oblong or elliptic leaflets 6-40 mm long and 2-14 mm wide, with sparse woolly hairs below. **Flowers** Pink-purple to almost white, pea-shaped, 15-25 mm long with a notched orbicular standard petal, arranged in axillary racemes of 12-35 flowers. **Fruits** Elliptic pods, 3-5 cm long. **Flowering** Spring and early summer. **Habitat** Heavy, grey soils along the banks and flats of the lower Murray River and the Darling River and its major tributaries in Qld, NSW, Vic. and SA. **Family** Fabaceae.

Swainsona lessertiifolia
Coast Swainson Pea

Upright or sprawling perennial shrubby herb with stout, silky-hairy stems to 30 cm long, poisonous to stock. **Leaves** Alternate, pinnately divided into 9-21 oblong-elliptic leaflets 5-25 mm long and 2-10 mm wide, with soft hairs below. **Flowers** Bright purple, rarely white, pea-shaped, 8-10 mm long and 12-14 mm across with a notched orbicular standard petal, arranged in axillary racemes of 12-35 flowers on stiff stalks. **Fruits** Pea-like pods, 15-30 mm long. **Flowering** Winter, spring and early summer. **Habitat** Widespread on coastal sandy sites in Vic., SA and Tas. **Family** Fabaceae.

Swainsona phacoides
Dwarf Swainson Pea

Variable prostrate or ascending perennial herb to 60 cm high with stout, silky-hairy stems, poisonous to stock. **Leaves** 5-10 cm long, pinnately divided into 5-13 narrow to broad elliptic leaflets, 5-30 mm long and 1-5 mm wide, flat or concave with soft hairs on both sides. **Flowers** Pale-violet to dark reddish purple, rarely white or yellow, pea-shaped, 10-15 mm long with a notched orbicular standard petal, arranged in axillary racemes of 1-10 flowers on hairy stalks. **Fruits** Oblong-elliptic hairy pods, 10-35 mm long. **Flowering** From spring to autumn. **Habitat** Arid, sandy sites in all mainland states. **Family** Fabaceae.

Swainsona swainsonioides
Downy Swainson Pea

Variable erect or spreading perennial herb to 60 cm high, often forming clumps, with white silky-hairy stems, poisonous to stock. **Leaves** 3-12 cm long, pinnately divided into 11-21 obovate, elliptical to oblong leaflets, silky hairy, 5-20 mm long and 2-12 mm wide. **Flowers** Blue to purple, pea-shaped, 8-15 mm long with a notched orbicular standard petal and coiled keel, arranged in axillary racemes of 6-15 flowers on stout stalks. **Fruits** Silky-hairy, oblong-elliptic pods, 10-35 mm long. **Flowering** Winter and spring. **Habitat** Clay and loam soils in woodlands and grasslands inland in southern Qld, NSW, western Vic. and SA. **Family** Fabaceae.

Swainsona greyana

Swainsona swainsonioides

Swainsona phacoides

Glycine clandestina

Swainsona lessertiifolia

Swainsona galegifolia

Kennedia beckxiana

Prostrate perennial herb with woody, twining stems covered with silky white hairs. **Leaves** Alternate, divided into 3 oblong to ovate leaflets with wavy margins, 1-2 cm long, sometimes covered with silky hairs. **Flowers** Red with yellow markings, pea-shaped, about 3 cm long with an orbicular standard petal and long keel, solitary or in axillary pairs on long stalks. **Fruits** Hairy flat pods. **Flowering** Spring. **Habitat** Moist situations in coastal southern WA. **Family** Fabaceae.

Kennedia prostrata   Running Postman. Scarlet Coral Pea

Prostrate perennial herb with twining stems covered with silky white hairs. **Leaves** Alternate, pinnately divided into 3 orbicular to broad obovate leaflets often with wavy margins, 5-30 mm long and 15-20 mm wide, covered with silky hairs. **Flowers** Scarlet with a greenish-yellow centre, pea-shaped, 15-20 mm long with a broad obovate standard petal and long keel, arranged in short axillary racemes of 1-4 flowers. **Fruits** Hairy, brown, cylindrical pods 3-5 cm long. **Flowering** Winter and spring. **Habitat** Sandy soils and rocky outcrops mainly along the coast or inland districts of NSW, Vic., SA and northeastern Tas. **Family** Fabaceae.

Kennedia rubicunda   Red Bean. Dusky Coral Pea

Prostrate perennial herb with twining stems to 4 m long, usually covered with brown hairs. **Leaves** Alternate, pinnately divided into 3 ovate to lanceolate leaflets, 3-15 cm long and 15-80 mm wide, covered with silky hairs, on long, slender stalks. **Flowers** Dark-red or purple with a brownish centre, pea-shaped, 25-40 mm long with a narrow obovate standard petal and long keel, arranged in axillary racemes of 2-12 flowers, 4-10 cm long. **Fruits** Hairy flat linear pods, 5-10 cm long. **Flowering** Late winter and spring. **Habitat** Widespread in a variety of habitats, particularly on sandy soils of the coast and tablelands in Qld, NSW, southeastern Vic. and northeastern Tas. **Family** Fabaceae.

Swainsonia formosa (syn. Clianthus formosus)   Sturt's Desert Pea

The South Australian floral emblem, a prostrate, spreading, annual or perennial herb with thick upright flowering stems to 1 m long. **Leaves** 10-15 cm long, pinnately divided into 11-17 narrow to broad elliptic or oblong leaflets, 1-3 cm long and 5-15 mm wide, grey-green and covered with long, soft hairs below. **Flowers** Bright red, white or a combination of both pea-shaped, 5-6 cm long with a pointed ovate standard petal, usually with a large, raised, black, glossy spot at the base, and a long curved keel, arranged in dense pendulous axillary racemes of 2-6 flowers. **Fruits** Narrow-elliptic swollen hairy pods, 4-9 cm long. **Flowering** Spring. **Habitat** Sandy soils in arid inland areas in open sites or mulga woodlands of all mainland states except Vic. **Family** Fabaceae.

Bossiaea walkeri   Cactus Bossiaea. Cactus Pea

Erect or spreading, rigid shrub to 2 m high with flattened grey branches, often with a waxy surface. **Leaves** Reduced to elliptical flattened scales, 15-25 mm long and to 8 mm wide. **Flowers** Bright red, pea-shaped, 2-3 cm long with an orbicular standard petal and longer, narrow keel, solitary on short axillary stalks. **Fruits** Flat, oblong pods about 6 cm long and 1 cm wide. **Flowering** Autumn, winter and spring. **Habitat** Semi-arid areas in mallee scrub and low open forests in western NSW, SA and southwestern WA. **Family** Fabaceae.

Templetonia retusa   Parrot Bush. Red Templetonia

Erect, occasionally prostrate shrub to 4 m high with rigid branches. **Leaves** Alternate, leathery, oblong to obovate or wedge-shaped, often with a notched tip, grey-green, 2-4 cm long and 6-25 mm wide. **Flowers** Deep red, rarely yellow, pea-shaped, 3-5 cm long with a curled-back standard petal and long, narrow keel, arranged in small terminal or axillary clusters. **Fruits** Flat oblong pods, 4-8 cm long and 10-16 mm wide. **Flowering** Winter and spring. **Habitat** Sandy soils and rocky slopes in low open forests and mallee scrub in dry coastal and inland areas of SA and southwestern WA. **Family** Fabaceae.

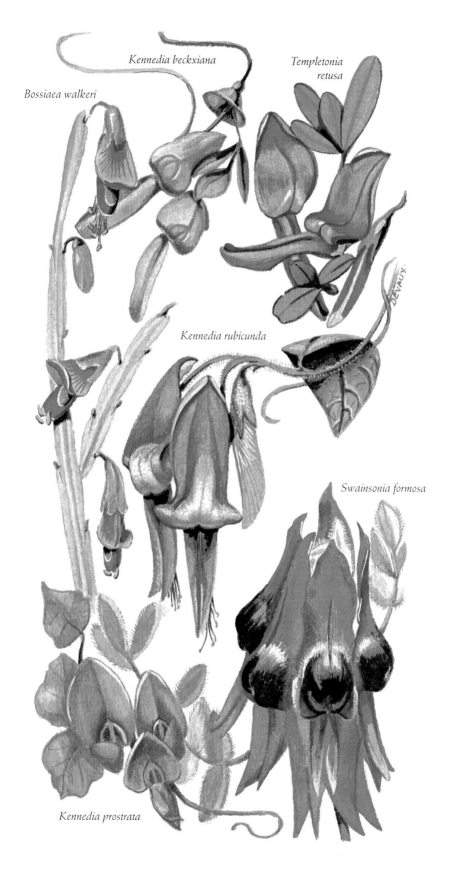

Bossiaea walkeri

Kennedia beckxiana

Templetonia retusa

Kennedia rubicunda

Swainsonia formosa

Kennedia prostrata

Swainsona stipularis

Erect or semi-prostrate perennial herb to 40 cm high, covered with short white hairs, poisonous to stock. **Leaves** Alternate, 3-9 cm long, pinnately divided into 5-11 lanceolate to oblong or wedge-shaped leaflets, notched at the tip, 1-25 mm long and 1-4 mm wide, both surfaces covered with silky hairs. **Flowers** Reddish-brown to pale purple, pea-shaped, 7-15 mm long with a notched, orbicular standard petal and slightly twisted keel, arranged in axillary racemes of 5-20 flowers on long, hairy stalks. **Fruits** Cylindrical pods covered in soft hairs, 1-3 cm long. **Flowering** Winter and spring. **Habitat** Widespread in dry inland areas of southern Qld, NSW, western Vic., SA and WA. **Family** Fabaceae.

Oxylobium alpestre   **Alpine Oxylobium. Alpine Shaggy Pea**

Low shrub to 1.5 m high with downy young stems. **Leaves** Opposite or in whorls of 3, oblong to narrow-ovate with curved-back margins and a depressed midrib, leathery, 1-4 cm long and 3-12 mm wide. **Flowers** Yellow-orange with a red centre, pea-shaped, 10-12 mm long, with a broad notched standard petal, arranged in short terminal or axillary racemes. **Fruits** Ovoid to oblong swollen pods to 15 mm long. **Flowering** Summer. **Habitat** Higher montane woodlands, often on rocky sites in the southern tablelands of NSW and southeastern Vic. **Family** Fabaceae.

Oxylobium procumbens   **Trailing Oxylobium. Trailing Shaggy Pea**

Prostrate shrub with ascending stems to 30 cm high, hairy when young. **Leaves** Opposite or in threes, heart-shaped to ovate with a sharp point, rigid and leathery with prominent veins, 10-25 mm long and 6-12 mm wide. **Flowers** Yellow or orange and red, pea-shaped, 10-14 mm long with a broad notched standard petal, arranged in loose terminal or axillary racemes. **Fruits** Ovoid to oblong hairy swollen pods 12-15 mm long. **Flowering** Spring and summer. **Habitat** Sclerophyll forests and woodlands in the southern tablelands of NSW and Vic. **Family** Fabaceae.

Chorizema cordatum   **Heart-leaved Flame Pea**

Scrambling or climbing shrub to 2 m high with many slender weak branches. **Leaves** Alternate, heart-shaped to ovate or lanceolate, sometimes with toothed or lobed margins and conspicuous veins, rough, 2-5 cm long. **Flowers** Orange-yellow to red, pea-shaped, about 15 mm across with a kidney-shaped standard petal and short keel, arranged in loose terminal racemes on short stalks. **Fruits** Swollen pods. **Flowering** Late winter and spring. **Habitat** Sandy and gravelly sites in forests of southwestern WA. **Family** Fabaceae.

Dillwynia hispida   **Red Parrot Pea**

Erect or spreading wiry shrub to 60 cm high with hairy branches. **Leaves** Alternate, narrow-linear to needle-like, usually with stiff, spreading hairs, 3-12 mm long. **Flowers** Red to orange-red, pea-shaped, 7-12 mm long with a kidney-shaped standard petal usually with a yellow centre, arranged in loose terminal clusters of 1-5 flowers. **Fruits** Swollen pods. **Flowering** Late winter and spring. **Habitat** Mallee communities on sand or red soil in southwestern NSW, western Vic. and SA. **Family** Fabaceae.

Pultenaea elliptica

Compact, often dense, hairy, erect shrub to 1.5 m high. **Leaves** Crowded, mostly alternate, concave, oblong-elliptic, fleshy, hairy when young, 7-17 mm long and 2-6 mm wide. **Flowers** Orange-yellow, pea-shaped, 10-15 mm long with an orbicular standard petal usually with a yellowish centre and long keel, arranged in dense leafy racemes, almost stalkless. **Fruits** Ovate pods about 5 mm long. **Flowering** Summer. **Habitat** Widespread usually in dry sclerophyll forests and sandy heaths of the coast and tablelands in NSW. **Family** Fabaceae.

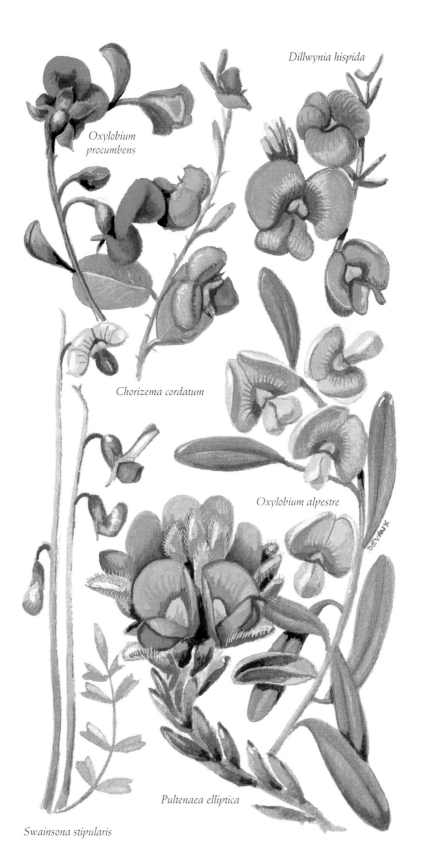

Dillwynia hispida

Oxylobium procumbens

Chorizema cordatum

Oxylobium alpestre

Pultenaea elliptica

Swainsona stipularis

Oxylobium ilicifolium — Prickly Oxylobium. Prickly Shaggy Pea

Erect shrub to 3 m high. **Leaves** Mostly opposite, 2-10 cm long and 1-3 cm wide, with irregularly-lobed margins, sometimes divided into 3 or more ovate lobes with sharply-pointed tips, dark-green above, stiff with prominent veins. **Flowers** Yellow-orange often with a red centre, pea-shaped, 8-12 mm long with a broad notched standard petal, arranged in short terminal or axillary racemes. **Fruits** Narrow-oblong swollen, hairy pods, 1-2 cm long and 2-3 mm wide. **Flowering** Spring and early summer. **Habitat** Widespread in sclerophyll forests on poor, shallow soils along the coast and tablelands in southeastern Qld, NSW and southeastern Vic. **Family** Fabaceae.

Jacksonia scoparia — Dogwood

Spreading shrub to small tree to 4 m high with grey-green pendulous angular branches and dark bark. **Leaves** Reduced to minute scales. **Flowers** Cream to orange-yellow, pea-shaped, 5-10 mm long with a notched orbicular standard petal and silky-hairy calyx, arranged in long terminal racemes. **Fruits** Flat, hairy, oblong pods, 6-12 mm long. **Flowering** Spring and early summer. **Habitat** Widespread on poor, gravelly soils in many situations on hillsides and ridges along the coast, tablelands and western slopes of southeastern Qld and NSW. **Family** Fabaceae.

Daviesia latifolia — Hop Bitter Pea

Erect shrub to 3 or rarely 5 m high with long flexible stems. **Leaves** (phyllodes) Alternate, ovate to lanceolate, rigid with prominent veins and usually with wavy margins, 2-14 cm long and 5-50 mm wide. **Flowers** Yellow with a dark-red centre, pea-shaped, 4-7 mm long with a notched orbicular standard petal, arranged in long dense axillary racemes. **Fruits** Compressed triangular pods 7-9 mm long and 4-7 mm wide. **Flowering** Spring and early summer. **Habitat** Widespread in sandy soils, especially in dry sclerophyll forests to 1800 m along the coast and ranges of southeastern Qld, NSW, Vic. and northeastern Tas. **Family** Fabaceae.

Daviesia mimosoides — Narrow-leaf Bitter Pea

Erect shrub to 2 m high, rarely a small tree to 5 m. **Leaves** (phyllodes) Alternate, lanceolate to ovate or spathulate with pointed tips, thick and rigid, 2-20 cm long and 4-30 mm wide. **Flowers** Yellow with a dark-red centre, pea-shaped, 5-7 mm across with a notched orbicular standard petal, arranged in dense axillary or terminal racemes 1-3 cm long, with 5-10 flowers. **Fruits** Triangular pods 7-10 mm long and 4-7 mm wide. **Flowering** Spring. **Habitat** Widespread in shallow soils on hillsides in sclerophyll forests to 1500 m along the coast and ranges of southeastern Qld, NSW and eastern Vic. **Family** Fabaceae.

Daviesia ulicifolia — Gorse Bitter Pea. Native Gorse

Erect shrub to 2 m high with many short angular rigid branches usually ending in spines. **Leaves** (phyllodes) Alternate, ovate to narrow-elliptic, rigid with long sharp points, sometimes concave, 5-20 mm long and 1-6 mm wide. **Flowers** Yellow-orange with a red centre, pea-shaped, 5-7 mm long with a notched orbicular standard petal, solitary or in axillary clusters of 2-4 flowers. **Fruits** Compressed triangular pods 5-8 mm long and 3-5 mm wide. **Flowering** Late winter and spring. **Habitat** Widespread in sandy and rocky sites in dry sclerophyll forests along the coast, ranges and inland in Qld, NSW, Vic., SA, western WA and northeastern Tas. **Family** Fabaceae.

Daviesia leptophylla (syn. D. virgata) — Slender Bitter Pea

Erect shrub to 2 m high with grey-green ridged branches. **Leaves** (phyllodes) Alternate, narrow-ovate to linear, thick and rigid with longitudinal veins, dull yellow-green, 1-10 cm long and 1-10 mm wide. **Flowers** Yellow to brown with a red centre, pea-shaped, 5-7 mm long with a notched orbicular standard petal, clustered in short racemes of 5-10 flowers. **Fruits** Triangular pods, 4-10 mm long and 4-6 mm wide. **Flowering** Spring and early summer. **Habitat** Open forests to dry, open sites on shallow soils of the coast and tablelands of NSW, Vic. and SA. **Family** Fabaceae.

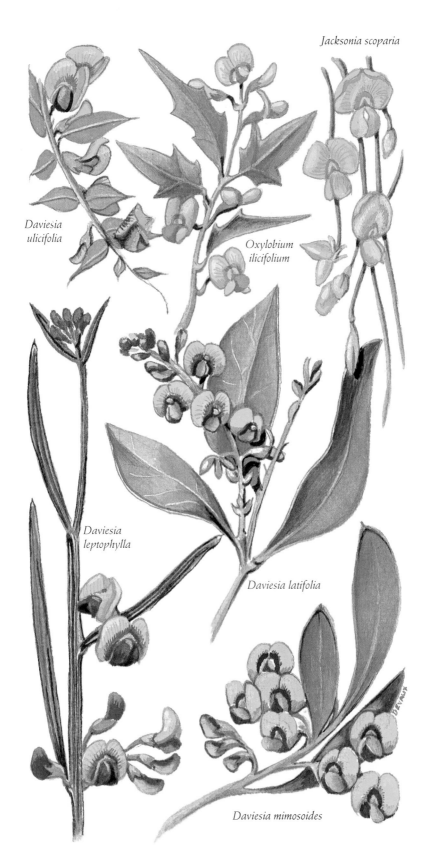

Jacksonia scoparia

Daviesia
ulicifolia

Oxylobium
ilicifolium

Daviesia
leptophylla

Daviesia latifolia

Daviesia mimosoides

Pultenaea cunninghamii
Grey Bush Pea

Spreading shrub to 2 m high with graceful drooping branchlets. **Leaves** Opposite or in whorls of 3, broad-ovate to broad elliptic with a sharp point and prominent mid-vein, grey-green, 6-22 mm long and 4-25 mm wide. **Flowers** Orange-yellow with red markings and keel, pea-shaped, 8-15 mm long with a folded notched orbicular standard petal, arranged in small axillary clusters on stalks 2-5 mm long. **Fruits** Ovate pods 6-7 mm long. **Flowering** Winter, spring and early summer. **Habitat** Poor soils in open forests on drier rocky hillsides of the coast, tablelands and inland slopes of southeastern Qld, NSW and Vic. **Family** Fabaceae.

Urodon dasyphylla *(syn. Pultenaea dasyphylla)*
Mop Bush Pea

Prostrate spreading shrub to 30 cm high, forming large mats. **Leaves** Crowded, very hairy, lanceolate, often grey, 6-10 mm long and 2-4 mm wide. **Flowers** Orange-yellow with red markings, pea-shaped, about 12 mm across with an orbicular standard petal and hairy calyx, arranged in dense racemes. **Fruits** Small ovoid pointed hairy pods, 5-7 mm long. **Flowering** Spring. **Habitat** Sandy heaths from central western to southwestern WA. **Family** Fabaceae.

Pultenaea mollis
Narrow-leaf Bush Pea

Erect bushy shrub to 2.5 m high with soft hairy branchlets. **Leaves** Crowded, upcurving, narrow, needle-like or concave with soft hairs, 8-20 mm long. **Flowers** Yellow with red markings, pea-shaped, about 1 cm across with a notched orbicular standard petal, almost stalkless in small terminal clusters. **Fruits** Small ovate pods. **Flowering** Spring. **Habitat** Open forests in Vic. **Family** Fabaceae.

Pultenaea polifolia

Weak shrub with spreading or ascending branches to 1 m long. **Leaves** Alternate, linear to narrow-elliptic with curved-back margins, sometimes pointed, silky-hairy below, 2-40 mm long and 2-7 mm wide. **Flowers** Yellow with red markings, pea-shaped, 5-10 mm long with a notched orbicular standard petal and hairy calyx, arranged in small terminal clusters. **Fruits** Flattened ovate pods, 5-10 mm long. **Flowering** Spring. **Habitat** Widespread in open forests and heaths in wet and dry sites along the coast and ranges of NSW and southeastern Vic. **Family** Fabaceae.

Pultenaea scabra
Rough Bush Pea

Erect shrub to 2 m high with hairy stems. **Leaves** Alternate, narrow to broad wedge-shaped, usually notched at the tip with curved-back margins and a depressed midrib, rough and stiff with short hairs above, paler green below, 3-16 mm long and 2-13 mm wide. **Flowers** Yellow with red markings, pea-shaped, 7-12 mm long with a notched orbicular standard petal, arranged in small terminal or axillary clusters of 2-5 flowers on very short stalks. **Fruits** Brown ovate, flattened pods 5-7 mm long. **Flowering** Spring. **Habitat** Widespread in open forests and heaths, usually on sandy soils of the coast and ranges of NSW, Vic. and southeastern SA. **Family** Fabaceae.

Pultenaea stipularis

Erect, stiff shrub to 2 m high. **Leaves** Alternate, crowded, linear to narrow-elliptic, pointed and rigid, 15-40 mm long and 1-2 mm wide. **Flowers** Yellow, often with small red markings, pea-shaped, 10-15 mm long with a notched folded orbicular standard petal, stalkless in dense terminal heads. **Fruits** Ovate pods, 6-7 mm long. **Flowering** Spring. **Habitat** Widespread on sandy soils in dry sclerophyll forests in sheltered sites along the central and southern coast of NSW. **Family** Fabaceae.

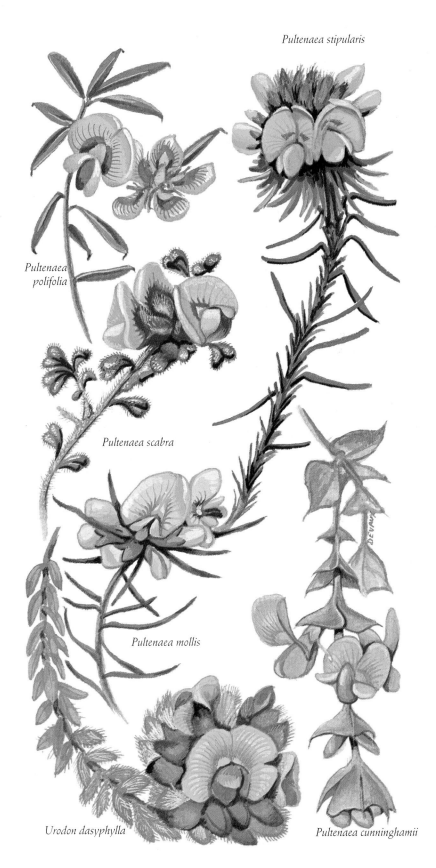

Pultenaea stipularis

Pultenaea polifolia

Pultenaea scabra

Pultenaea mollis

Urodon dasyphylla

Pultenaea cunninghamii

Pultenaea daphnoides
Large-leaf Bush Pea

Erect shrub to 3 m high with downy stems. **Leaves** Alternate, oblong to broad wedge-shaped with a pointed tip, paler below with a prominent midrib, 5-40 mm long and 2-13 mm wide. **Flowers** Yellow with red markings, pea-shaped, 7-15 mm long with a slightly notched folded orbicular standard petal and silky-hairy calyx, arranged in dense terminal heads 2-3 cm across. **Fruits** Small flattened ovate pods, 5-7 mm long. **Flowering** Spring and summer. **Habitat** Widespread in sandy soils in heaths and sclerophyll forests along the coast and ranges of southeastern Qld, NSW, Vic., SA and Tas. **Family** Fabaceae.

Pultenaea hispidula
Rusty or Scented Bush Pea

Erect shrub to 2 m high with drooping, hairy branches. **Leaves** Crowded, oblong, concave, hairy and leathery, 3-10 mm long and 1-2 mm wide. **Flowers** Yellow with red markings, pea-shaped, 5-8 mm long with a folded orbicular standard petal, arranged in loose, leafy, terminal heads. **Fruits** Brown ovate pods, 4-5 mm long. **Flowering** Spring. **Habitat** Scattered in sandy soils of heaths and dry sclerophyll forests of the central and southern coast and tablelands of NSW, Vic. and eastern SA. **Family** Fabaceae.

Pultenaea juniperina
Prickly Bush Pea

Erect or straggling prickly shrub to 4 m high. **Leaves** Alternate, usually concave, linear to narrow-elliptic with a pointed tip, stiff, 7-30 mm long and 1-3 mm wide. **Flowers** Yellow with red markings, pea-shaped, 7-13 mm long with a notched folded orbicular standard petal, solitary or 2-3 flowers in terminal or axillary clusters. **Fruits** Small flattened ovate pods, 5-8 mm long. **Flowering** Spring. **Habitat** Widespread in wet sclerophyll forests in the northern and southern tablelands of NSW, Vic. and Tas. **Family** Fabaceae.

Pultenaea villosa
Hairy Bush Pea

Erect or prostrate shrub with hairy branches to 3 m long. **Leaves** Alternate, narrow-oblong or slightly obovate with curved-back margins, hairy below, 3-10 mm long and 1-3 mm wide. **Flowers** Yellow with red markings, pea-shaped, 5-12 mm long with a notched folded orbicular standard petal, crowded in the leaf axils, or in terminal heads. **Fruits** Ovate pods, 4-6 mm long. **Flowering** Most of the year. **Habitat** Widespread from coastal heaths to tall forests in southeastern Qld and eastern NSW. **Family** Fabaceae.

Mirbelia oxylobioides
Mountain Mirbelia

Erect or spreading wiry shrub to 1.5 m high. **Leaves** Opposite, scattered or in whorls of 3, elliptic to narrow-ovate, 2-10 mm long and 1-3 mm wide, with curved-back margins and depressed midrib, stiff, dark-green above and paler below with short, flattened hairs. **Flowers** Orange-yellow with red markings, pea-shaped, 8-10 mm long with a notched orbicular standard petal, in small terminal racemes. **Fruits** Hairy longitudinally-grooved compressed ovoid pods, 8-10 mm long. **Flowering** Summer. **Habitat** Open forests, mainly at higher altitudes in eastern NSW and Vic. **Family** Fabaceae.

Phyllota phylicoides
Heath Phyllota

Erect, spreading shrub to 1.5 m high often with reddish stems. **Leaves** Crowded, scattered, linear, almost stalkless with curled-under margins, rough, yellow-green, 5-20 mm long and 1-2 mm wide. **Flowers** Yellow to orange with a red calyx, pea-shaped, 5-12 mm long with a broad folded standard petal, almost stalkless in compact, leafy, terminal spikes. **Fruits** Ovate to oblong pods about 5 mm long with 2 seeds. **Flowering** Spring. **Habitat** Widespread on sandy soils of heaths, scrubs and sclerophyll forests along the coast and ranges of southeastern Qld and NSW. **Family** Fabaceae.

Mirbelia oxylobioides

Pultenaea juniperina

Pultenaea hispidula

Phyllota phylicoides

Pultenaea villosa

Pultenaea daphnoides

Bossiaea cinerea   Showy Bossiaea

Erect or straggling shrub to 1 m high with stiff, wiry, hairy branches. **Leaves** Alternate, opposite or whorled, stalkless, narrow-ovate to almost triangular with a pointed tip and curved-back margins, paler and hairy below, 5-20 mm long and 3-6 mm wide. **Flowers** Yellow with reddish brown markings, pea-shaped, 7-15 mm long with a notched orbicular standard petal and long narrow keel, solitary on fine axillary stalks 4-10 mm long. **Fruits** Flat oblong pods, 1-2 cm long. **Flowering** Late winter and spring. **Habitat** Coastal heaths and light forests in southern NSW, Vic., southeastern SA and Tas. **Family** Fabaceae.

Bossiaea ensata   Sword Bossiaea

Erect or straggling shrub to 1.5 m high with flat, leathery branches. **Leaves** Reduced to scales, 1-2 mm long, developing on seedlings and new growth after fire. **Flowers** Yellow with dark-red markings and red outside, pea-shaped, 6-10 mm long with a notched orbicular standard petal and narrow keel, solitary on axillary stalks 2-5 mm long. **Fruits** Flat 2-valved, oblong pods, 25-40 mm long and 1 cm wide. **Flowering** Late winter and spring. **Habitat** Sandy soils of coastal heaths and light forests in southeastern Qld, NSW, southeastern Vic. and the Eyre Peninsula in SA. **Family** Fabaceae.

Bossiaea foliosa   Leafy Bossiaea

Erect, dense shrub to 1.5 m high with stiff, wiry branches. **Leaves** Alternate, crowded, small and concave, orbicular, shiny dark-green above and sometimes hairy below, 2-4 mm wide. **Flowers** Yellow, pea-shaped, 5-7 mm long with a notched orbicular standard petal and long narrow keel, solitary on short axillary stalks, forming dense clusters. **Fruits** Flat, circular, hairy pods, 5-10 mm diameter, covered with thick brown hair. **Flowering** Spring and summer. **Habitat** At high altitude on rocky plateaus and near streams in montane forests of central and southern NSW and southeastern Vic. **Family** Fabaceae.

Bossiaea heterophylla

Erect shrub to 1.5 m high with flattened young stems. **Leaves** Alternate, stalkless, lanceolate to ovate, 6-30 mm long and 2-4 mm wide. **Flowers** Yellow with a red keel, pea-shaped, 7-15 mm long with a notched orbicular standard petal and long narrow keel, solitary on short axillary stalks. **Fruits** Flat oblong pods 2-4 cm long and 7 mm wide. **Flowering** Mostly in autumn. **Habitat** Common on sandy soils in a variety of habitats, particularly coastal heaths and light forests of southeastern Qld, eastern NSW and southeastern Vic. **Family** Fabaceae.

Bossiaea scolopendria

Erect shrub to 1.5 m high with flattened stems 6-15 mm wide. **Leaves** Reduced to scales 1-2 mm long only on seedlings and new growth after fire. **Flowers** Yellow with a red keel, pea-shaped, 10-15 mm long with a notched orbicular standard petal and narrow keel, solitary on short stalks arising from the stems. **Fruits** Flat oblong pods 3-5 cm long and 1 cm wide. **Flowering** Spring and early summer. **Habitat** Widespread in heaths and dry sclerophyll forests of central and southern coastal NSW. **Family** Fabaceae.

Daviesia corymbosa   Narrow-leaved Bitter Pea

Erect shrub to 2 m high. **Leaves** (phyllodes) Very variable, alternate, linear to narrow-ovate to lanceolate, rigid with prominent veins, 2-12 cm long and 2-25 mm wide. **Flowers** Yellow to orange with a dark-red centre, pea-shaped, 4-8 mm long with a notched orbicular standard petal, arranged in crowded axillary racemes of 5-20 flowers on stalks 1-3 cm long. **Fruits** Flattened triangular pods 8-9 mm long and about 6 mm wide. **Flowering** Spring. **Habitat** Widespread in coastal sandy heaths and dry sclerophyll forests along the coast and tablelands of NSW. **Family** Fabaceae.

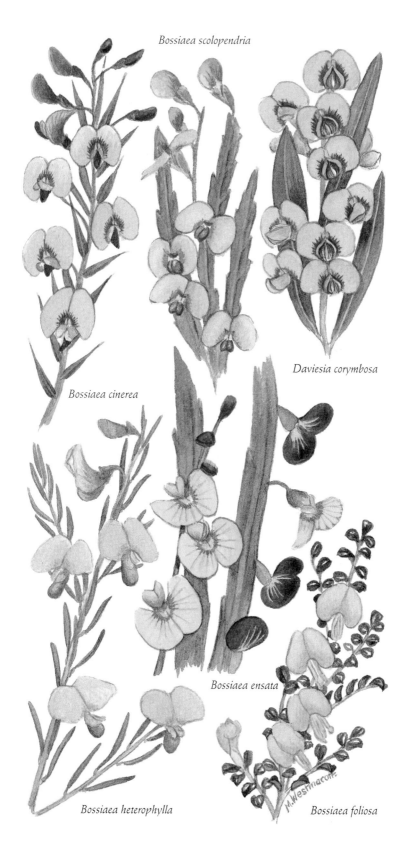

Bossiaea scolopendria

Daviesia corymbosa

Bossiaea cinerea

Bossiaea ensata

Bossiaea heterophylla

Bossiaea foliosa

Dillwynia glaberrima
Smooth Parrot Pea

Spreading wiry shrub to 2 m high with slender arching branches. **Leaves** Alternate, narrow-linear to needle-like, usually with a small hooked tip, 5-25 mm long. **Flowers** Yellow with a reddish brown centre, pea-shaped, 8-10 mm long with a kidney-shaped standard petal, usually 1-4 flowers on stalks 3-20 mm long in the upper leaf axils. **Fruits** Swollen ovoid hairy pods, 4-6 mm long. **Flowering** Spring. **Habitat** Widespread in sandy heaths and open forests in southeastern Qld, coastal NSW, Vic., southeastern SA and Tas. **Family** Fabaceae.

Dillwynia retorta

Erect or prostrate shrub to 3 m high with hairy branches. **Leaves** Alternate, narrow-linear and twisted, 4-12 long. **Flowers** Yellow with a red centre, pea-shaped, 4-12 mm long with a kidney-shaped standard petal, clustered at the ends of the branches, with or without stalks. **Fruits** Swollen pods, 4-7 mm long. **Flowering** Spring and early summer. **Habitat** Widespread in sandy heaths and dry sclerophyll forests in southeastern Qld and eastern NSW. **Family** Fabaceae.

Dillwynia sericea
Showy Parrot Pea

Bushy shrub to 1 m high with silky-hairy branches. **Leaves** Alternate, narrow-linear to needle-like, grooved above, sometimes slightly hairy, 5-20 mm long. **Flowers** Yellow or orange, often with a red centre, pea-shaped, 8-12 mm long with a kidney-shaped standard petal, solitary or in pairs in the upper leaf axils, forming a leafy spike. **Fruits** Swollen pods 3-4 mm long. **Flowering** Spring and early summer. **Habitat** Dry sclerophyll forests, woodlands and heaths along the coast, ranges and inland slopes of southeastern Qld, NSW, Vic., southeastern SA and Tas. **Family** Fabaceae.

Gompholobium latifolium
Giant Wedge Pea. Golden Glory Pea

Erect shrub to 3 m high with long slender stems. **Leaves** Alternate, divided into 3 linear to wedge-shaped leaflets with curved-back margins, pale-green below, 2-6 cm long and 2-6 mm wide. **Flowers** Yellow, pea-shaped, 2-3 cm long with a notched orbicular standard petal, solitary or 2-3 together in the upper leaf axils. **Fruits** Almost globular pods 15-18 mm long. **Flowering** Spring and early summer. **Habitat** Widespread on poor sandy soils in dry sclerophyll forests along the coast and ranges of southeastern Qld, NSW and Vic. **Family** Fabaceae.

Gompholobium virgatum
Wallum or Leafy Wedge Pea

Erect shrub to 1.5 m high. **Leaves** Alternate, divided into 3 linear to lanceolate or obovate leaflets usually with a curved-back tip, 1-3 cm long and 1-4 mm wide. **Flowers** Yellow, pea-shaped, 12-20 mm long with a notched orbicular standard petal, solitary or 2-3 together in the upper leaf axils. **Fruits** Almost globular pods about 1 cm long. **Flowering** Winter and spring. **Habitat** Widespread in sandy soils in heaths and dry sclerophyll forests of the coast tablelands and western slopes in southeastern Qld, and NSW. **Family** Fabaceae.

Platylobium formosum
Handsome Flat Pea

Erect, straggling or prostrate shrub to 2.5 m high with wiry stems. **Leaves** Opposite, almost stalkless, ovate to lanceolate or heart-shaped, stiff and rough with raised veins, dark-green above, paler below, 10-65 mm long and 15-40 mm wide. **Flowers** Yellow with a red centre, pea-shaped, 8-15 mm long with a kidney-shaped standard petal and long keel, solitary or rarely 2 or 4 together on hairy, axillary stalks, 1-3 cm long. **Fruits** Flat, oblong, smooth or hairy pods, 2-4 cm long. **Flowering** Autumn, late winter and spring. **Habitat** Widespread in various habitats from heaths to rainforest margins of the coast and adjacent ranges in southeastern Qld, NSW, Vic. and Tas. **Family** Fabaceae.

Dillwynia glaberrima

Dillwynia retorta

Platylobium formosum

Dillwynia sericea

Gompholobium virgatum

Gompholobium latifolium

M. Westmacott.

Crotalaria cunninghamii
Green Bird Flower. Parrot Pea

Erect shrub to 2 m high with soft hairs on the branches. **Leaves** Alternate, entire or divided into 3 leaflets, ovate to elliptic, thick and hairy, 4-9 cm long and 1-6 cm wide. **Flowers** Yellow-green streaked with fine purple lines, pea-shaped, 3-6 cm long with an ovate pointed standard petal and long twisted keel, arranged in terminal racemes to 25 cm long. **Fruits** Swollen, oblong, hairy pods, 2-5 cm long and 10-12 mm wide. **Flowering** Autumn, winter and spring. **Habitat** Widespread on sandy soils inland in mulga communities and on sand dunes in Qld, NSW, SA, WA and NT. **Family** Fabaceae.

Crotalaria eremaea
Blue Bush Pea

Erect slender shrub to 2 m high with soft grey hairs on the branches. **Leaves** Alternate, entire or divided into 2-3 leaflets, linear to oblong or ovate, concave, grey-green, 2-6 cm long and 3-35 mm wide. **Flowers** Yellow, pea-shaped, 10-18 mm long with an ovate standard petal, arranged in loose terminal racemes, 8-40 cm long, of 15-30 flowers. **Fruits** Swollen, oblong, grey-hairy pods, 15-30 mm long and 3-7 mm wide. **Flowering** After rain, mostly in summer and autumn. **Habitat** Sandy soils inland in all mainland states except Vic. **Family** Fabaceae.

Crotalaria laburnifolia
Bird Flower

Erect shrub to 3 m high. **Leaves** Alternate, divided into 3 ovate to lanceolate leaflets on long stalks, greyish green, 25-50 mm long. **Flowers** Yellow-green, pea-shaped, 25-35 mm long with an ovate, pointed, standard petal and long beaked keel, arranged in long terminal racemes. **Fruits** Swollen pods, 3-4 cm long. **Flowering** Most of the year. **Habitat** Sandy soils in northeastern Qld and northern WA. **Family** Fabaceae.

Goodia lotifolia
Golden Tip

Erect open shrub to 4 m high. **Leaves** Alternate, divided into 3 ovate leaflets, paler below, 1-3 cm long and 5-8 mm wide. **Flowers** Yellow with red markings, pea-shaped, 9-13 mm long, with a notched orbicular standard petal, arranged in loose terminal racemes to 8 cm long. **Fruits** Flat oblong pods, 12-30 mm long and 6-8 mm wide. **Flowering** Winter and spring. **Habitat** Widespread pioneer species in disturbed along the coast and ranges of southeastern Qld, NSW, Vic., SA, southwestern WA and northeastern Tas. **Family** Fabaceae.

Sphaerolobium vimineum
Leafless Globe Pea

Erect wiry shrub to 70 cm high. **Leaves** Usually absent or reduced to a few small scales about 5 mm long. **Flowers** Yellow and orange, pea-shaped, 4-6 mm long with a notched orbicular standard petal and long keel, scattered along the branches in clusters of 2-3 flowers. **Fruits** Globular pods 3-5 mm diameter. **Flowering** Winter and spring. **Habitat** Widespread in sandy heaths and open forests, often in swamp sites of the coast and tablelands in southeastern Qld, NSW, Vic., SA, southwestern WA and Tas. **Family** Fabaceae.

Aotus ericoides
Common Aotus

Erect slender shrub to 2 m high, with woolly stems. **Leaves** Usually in whorls of 3, almost stalkless, linear to very narrow-ovate with curled-under margins, 6-20 mm long and 1-5 mm wide. **Flowers** Yellow, often with red markings in the centre, pea-shaped, 5-10 mm long with an orbicular standard petal and hairy calyx, solitary or in small axillary clusters, forming a leafy raceme. **Fruits** Hairy ovate pods, 4-7 mm long. **Flowering** Late winter and spring. **Habitat** Widespread in a wide range of habitats, particularly coastal heaths and dry sclerophyll forests in southeastern Qld, eastern NSW, Vic. and Tas. **Family** Fabaceae.

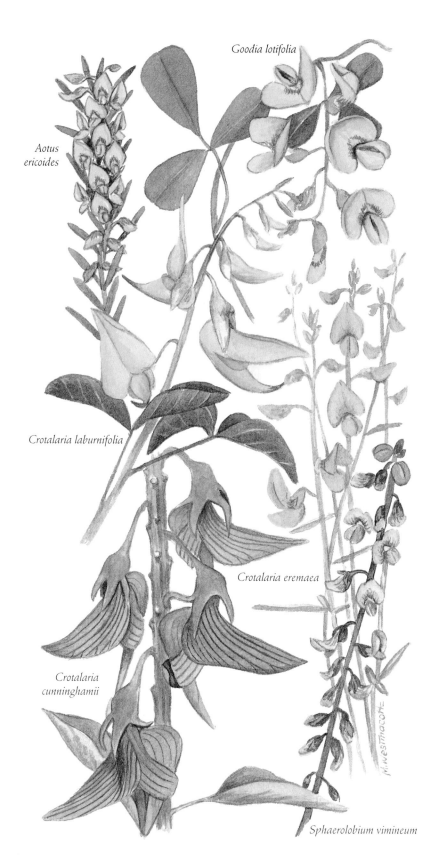

Goodia lotifolia

Aotus ericoides

Crotalaria laburnifolia

Crotalaria eremaea

Crotalaria cunninghamii

Sphaerolobium vimineum

Glossary of terms

achene a dry fruit with 1 seed, not splitting open when ripe.

alternate arranged one by one along a stem, not opposite.

annual completing its life cycle in one year.

anther the top end of the stamen, bearing pollen.

aquatic growing in water.

aromatic fragrant flowers or foliage.

ascending growing upwards.

axil the upper angle between leaf and stem or branch.

axillary arising from the axil.

bark outer covering of the stem or root.

beak pointed projection.

beard a tuft of hair.

belah Casuarina cristata, a dominant tree species in some inland areas.

berry succulent non-opening fruit, usually rounded usually with many seeds.

biennial completing its life cycle in 2 years.

bipinnate a leaf twice pinnately divided.

bract modified leaf often at the base of a flower or stem.

branchlet a small branch.

bristle short stiff hair.

bulb swollen underground stem acting as a storage organ.

calli small, hard protrusions.

calyx outer whorl of the flower, consisting of sepals.

capsule dry opening fruit of more than one carpel.

carpel female part of the flower usually comprising stigma, style and ovary.

clasping partly or wholly surrounding the stem.

column structure formed by fused stigmas and styles in orchids.

composite of the Compositae family, with many florets in a close head surrounded by a common whorl of bracts.

compound consisting of several similar parts.

compressed flattened.

cone a globular collection of fruits around a central axis, surrounded by woody bracts.

conical cone-shaped.

constricted drawn together, narrowed as between seeds in a pod.

creeping remaining close to the ground.

crown the leafy head of a tree.

cypsella a dry, single seeded fruit, not splitting open when ripe.

deciduous liable to be shed at a certain time.

decumbent lying on the ground with the tip turned up.

decussate leaves arranged opposite in pairs at right angles.

depressed flattened or sunken.

downy with short soft hairs.

drupe fleshy non-opening fruit with a hard kernel and solitary seed.

elliptical a plane surface shaped like an ellipse.

elongate extended in length.

entire undivided, without teeth or lobes.

epiphyte a plant growing on another plant or object, using it for support and not nourishment.

family a group of closely related genera.

filament stalk bearing the anther.

floral leaves leaves immediately below the flowers.

floret one of the small flowers in a compact head.

flower sexual reproductive structure.

follicle a dry fruit formed from 1 carpel and splitting open along the inner margin.

fruit seed-bearing part of a plant.

fused joined together.

genus a group of closely related species.

gland embedded or projecting structure usually secreting oil, nectar, resin or water.

globular globe-shaped, spherical or nearly so.

habitat natural abode of a plant.

head dense cluster of flowers or fruits.

heath an area occupied mainly by low, shrubby plants, whose growth is conditioned by severe environmental factors.

herb a plant without a woody stem.

inflorescence the flowering structure of a plant.

keel the two lower fused petals of a pea-like flower.

kino red or black juice or gum.

lanceolate lance-shaped, tapering at each end, broadest below the middle, about four times as long as broad.

lateral on the side or edge.

leaf usually a green flat organ attached to the stem, manufacturing food.

leaflet a secondary part of a compound leaf.

linear long and narrow.

lip one of the petals or sepals of a flower.

littoral near the sea.

lobe rounded or pointed division of a leaf; the sepal or petal of a flower.

mallee Eucalypts growing with several stunted stems, common in arid and alpine areas.

mangrove species of plants growing in salty water along coasts and estuaries.

margin edge.

membranous thin, flexible and sheet-like.

midrib main vein of a leaf, leaflet or segment, running from base to tip.

nut a dry non-opening fruit with one seed and hard woody covering.

oblong having roughly parallel sides, longer than broad with a rounded tip.

obovate almost ovate, but broader towards the tip.

opposite in pairs one at each side of the stem.

orbicular more or less circular in outline.

ovate egg-shaped, broadest below the middle.

ovary female structure enclosing the unfertilised seeds.

ovoid an egg-shaped solid body.

ovule he body in the ovary which becomes the seed after fertilisation.

palmate a leaf divided into three or more leaflets or lobes arising from a common point.

panicle a much-branched inflorescence.

perennial a plant living for more than two years.

petal a segment of the inner whorl of the floral lobes.

phyllode flattened leaf stalk resembling and acting as a leaf.

pinnate a compound leaf with leaflets on opposite side of a common leaf stalk.

pistil female reproductive organ in a flower.

pod dry, opening, multi-seeded fruit.

pollen powdery substance produced in the anthers.

prostrate lying on the ground.

prop roots roots growing down from a trunk or branch and supporting the tree.

raceme an inflorescence with stalked flowers borne along an unbranched axis.

rainforest a closed forest dominated by trees with soft leaves.

regular radially symmetrical.

rhizome a stem which is usually underground, producing new shoots and roots.

riverine situated beside a river.

scale a thin, dry, papery structure, a very small rudimentary leaf, or flat closely pressed leaf.

scattered leaves arranged in a random manner along the stem.

sclerophyll plants with harsh-textured, tough leaves.

scrub a community dominated by shrubs.

segment a subdivision of a divided or dissected leaf or other structure.

sepal a segment of the outer whorl of the flower.

serrate a leaf margin with many sharp teeth, as on a saw.

shrub a woody, perennial plant with several stems growing from the base, without a single trunk as in a tree.

silky covered with fine soft hair.

spathulate shaped like a spatula, tapering from a rounded tip to a narrow base.

species a group of individual plants essentially alike when grown under similar conditions, normally breeding freely with others of their own kind: the basic unit of biological classification.

spherical in the form of a globe.

spike a compact inflorescence of stalkless flowers.

spine
a sharp, rigid structure.

stamen male part of a flower comprising filament and anther.

stigma receptive tip of the style.

stipule one of a pair of scale or leaf-like appendages at the base of a leaf.

style stalk arising from the ovary and bearing the stigma.

succulent soft and juicy.

synonym (syn.) a plant name set aside in favour of an earlier one.

tepal petal or sepal, being scarcely distinguishable from each other.

terminal at the apex.

terrestrial plants growing in the ground, not aquatic or epiphytic.

tessellated in the form of small squares.

throat the opening of a flower tube.

tree a perennial plant with a single woody trunk and distinct head or crown.

trifoliolate having 3 leaflets.

tufted stems or leaves growing close together.

valve a cell or compartment in a fully matured capsule.

vein visible appearance of vascular tissue in a leaf.

whorl a group of three or more structures encircling an axis at the same level.

wing the membranous extension of a seed or fruit; the two lateral parts of a pea-shaped flower.

woolly having long, soft, matted hair.

Leaf shapes

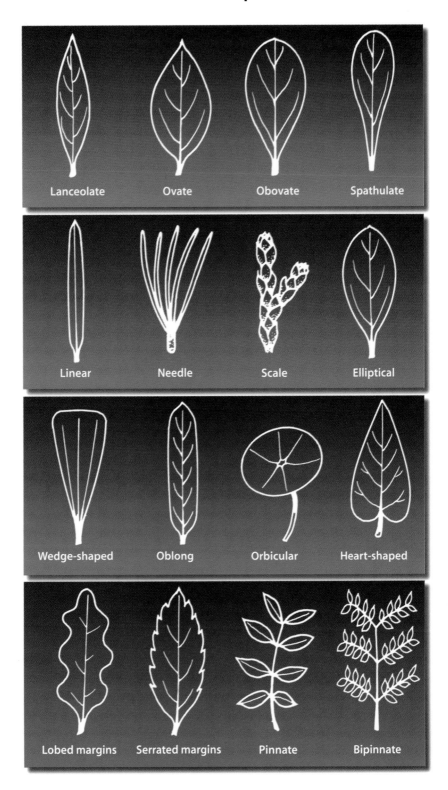

Lanceolate   Ovate   Obovate   Spathulate

Linear   Needle   Scale   Elliptical

Wedge-shaped   Oblong   Orbicular   Heart-shaped

Lobed margins   Serrated margins   Pinnate   Bipinnate

Flower parts and leaf arrangements

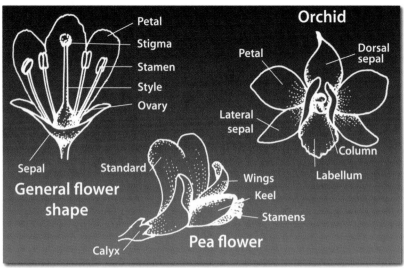

Petal
Stigma
Stamen
Style
Ovary

Sepal

Standard

General flower shape

Calyx

Orchid

Petal

Dorsal sepal

Lateral sepal

Column

Labellum

Wings
Keel
Stamens

Pea flower

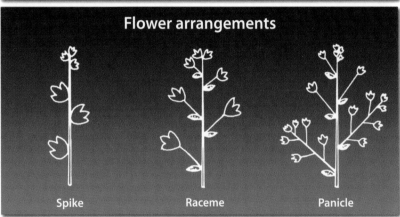

Flower arrangements

Spike

Raceme

Panicle

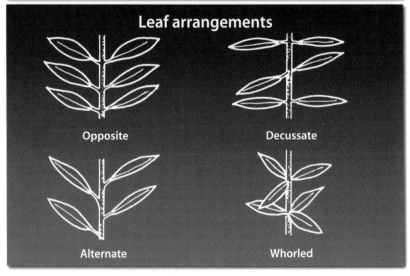

Leaf arrangements

Opposite

Decussate

Alternate

Whorled

Index